The OLYMPICS

Faye Cowin

Australia • Brazil • Japan • Korea • Mexico • Singapore • Spain • United Kingdom • United States

Maths in the Real World - The Olympics
1st Edition
Faye Cowin

Cover and text design: Cheryl Rowe
Production controllers: Jess Lovell & Siew Han Ong

Any URLs contained in this publication were checked for currency during the production process. Note, however, that the publisher cannot vouch for the ongoing currency of URLs.

First published in 1999 as Putting Maths to Work - The Olympics by New House Publishing.

Acknowledgements
Cover image courtesy of Shutterstock.
Images on pages 9, 11, 12, 13, 15, 16, 19, 21, 34, 39, 41 (top), 42, 55 (top left), 56 (right), 66 courtesy of Shutterstock.
Images and maps on pages 16 (map), 17, 35, 69 courtesy of London 2012.
Images on pages 30, 32, 38, 40, 41 (bottom), 44, 46, 48 (bottom), 51, 52, 53, 55 (bottom left and right), 56 (left), 64 courtesy of New Zealand Herald.
Stamps on page 62 courtesy of New Zealand Post Limited.

For product information and technology assistance,
in Australia call **1300 790 853**;
in New Zealand call **0800 449 725**

For permission to use material from this text or product, please email **aust.permissions@cengage.com**

National Library of New Zealand Cataloguing-in-Publication Data
Cowin, Faye.
The Olympics / Faye Cowin.
(Maths in the real world)
Previously published in the series: Putting mathematics to work. Auckland, N.Z. : New House Publishers, 1999.
ISBN 978-0-17-021709-5
1. Mathematics—Problems, exercises, etc. 2. Olympics—Problems, exercises, etc. 3. Sports—Problems, exercises, etc.
I. Title. II. Series: Cowin, Faye. Maths in the real world.
510.76—dc 22

Cengage Learning Australia
Level 7, 80 Dorcas Street
South Melbourne, Victoria Australia 3205

Cengage Learning New Zealand
Unit 4B Rosedale Office Park
331 Rosedale Road, Albany, North Shore 0632, NZ

For learning solutions, visit **cengage.com.au**

Printed in China by China Translation & Printing Services.
1 2 3 4 5 6 7 15 14 13 12 11

Contents

Introduction

Maths in the Real World is a series of theme based books identifying the necessary mathematical skills and knowledge needed in particular areas for the future. The books cover key aspects of the *New Zealand Maths and Statistics for New Zealand curriculum.* The skill requirements of the NCEA Level One achievement standards and remaining unit standards are covered, enabling students to be examined in these NCEA qualifications. They also address the requirements of the Numeracy Project.

Maths in the Real World is aimed at those students who wish to pursue a non-academic career but for whom mathematics is an essential component of their trade training or life in the future. The series is an alternative course in mathematics for 15 to 17 year- olds in schools or training centres. Together, the books offer students:

- Practice with basic mathematics and calculation skills, so essential in the transition from school to the next stage in their development – flatting, travelling, working, buying a car.
- Opportunities to apply mathematics in practical everyday situations – making budgets, shopping, earning money, paying bills, planning a trip, owning a car.
- An awareness of their individual rights and responsibilities and an introduction to the range of community facilities available to them.

Essential skills covered in this book

- Whole number and all operations
- Rounding of decimals
- Table reading
- Form filling
- Surveys
- Angles
- Problem solving
- Decimals and all operations
- Percentages
- Graph reading and drawing
- Measurement
- Transformations
- Scale drawings
- Constructions.

ISBN: 9780170217095

New Zealand curriculum level 5: Numbers and algebra

Number strategies and knowledge

- Reason with linear proportions.
- Use prime numbers, common factors and multiples and powers (including square roots).
- Understand operations on fractions, decimals, percentages and integers.
- Use rates and ratios.
- Know commonly used fraction, decimal and percentage conversions.
- Know and apply standard form, significant figures, rounding and decimal place value.

Measurement

- Select and use appropriate metric units for length, area, volume and capacity, weight (mass), temperature, angle and time, with awareness that measurements are approximate.
- Convert between metric units using decimals.
- Deduce and use formulae to find the perimeters and areas of polygons and the volumes of prisms.
- Find the perimeters and areas of circles and composite shapes and the volumes of prisms including cylinders.

Statistics

Statistical investigation

- Plan and conduct surveys and experiments using the statistical enquiry cycle:
  - determining appropriate variables and measures
  - considering sources of variation
  - gathering and cleaning data
  - using multiple displays, and recategorising data to find patterns, variations, relationships and trends in multivariate data sets.
- Comparing sample distributions visually, using measures of centre, spread and proportion.
- Presenting a report of findings.

Statistical literacy

- Evaluate statistical investigations or probability activities undertaken by others, including:
  - data collection
  - methods
  - choice of measures
  - validity of findings.

ISBN: 9780170217095

Literacy and Numeracy Standards for NCEA Level One

NCEA Level One Numeracy Standard	Number	Credits	New Zealand Curriculum
Apply numeric reasoning when solving problems	91026	4	NA6.1: Apply direct and inverse relationships with linear proportions NA6.2: Extend powers to include integers and fractions NA6.3: Apply everyday compounding rates
Solve measurement problems	91030	3	GM6.2: Apply the relationships between units in the metric system, including the units for measuring different attributes and derived measures GM6.3: Calculate volumes, including prisms, pyramids, cones and spheres, using formulae
Solve measurement problems involving right angled triangles	91032	3	GM6.1: Measure at a level of precision appropriate to the task GM6.5: Recognise when shapes are similar and use proportional reasoning to find an unknown length GM6.6: Use trigonometric ratios and Pythagoras' theorem in two and three dimensions
Apply transformation geometry	91034	2	GM6.8: Compare and apply single and multiple transformations GM6.9: Analyse symmetrical patterns by the transformations used to create them.
Use the statistical enquiry cycle to investigate bivariate numerical data	91036	3	S6.1: Plan and conduct investigations using the statistical enquiry cycle
Solve problems which require calculation with whole numbers	8489	2	Solve problems which require calculation with whole numbers (Expires Dec 2012)
Solve problems using calculations with numbers expressed in different forms	8490	2	Solve problems using calculations with numbers expressed in different forms (Expires Dec 2012)
Read and interpret information presented in tables and graphs	8491	2	Read and interpret information presented in tables and graphs (Expires Dec 2012)
Use measurement devices to measure quantities	8492	3	Use measurement devices to measure quantities (Expires Dec 2012)
Find relationships between measurements	12319	2	Find relationships between measurements (Expires Dec 2012)
Make estimates of measurements with common units	20662	2	Make estimates of measurements with common units (Expires Dec 2012)
Use numeracy strategies to solve problems involving whole numbers	23738	2	Use numeracy strategies to solve problems involving whole numbers (Expires Dec 2012)
Use numeracy strategies to solve number problems involving decimals, percentages and fractions	23739	2	Use numeracy strategies to solve number problems involving decimals, percentages and fractions (Expires Dec 2012)

ISBN: 9780170217095

These are the skills covered	New Zealand Curriculum - Level 5
• Whole number and all operations • Decimals and all operations • Rounding of decimals • Percentages • Table reading • Graph reading and drawing • Form filling • Measurement • Surveys • Transformations • Angles • Scale drawings • Problem-solving • Constructions	Number strategies and knowledge • Reason with linear proportions. • Use prime numbers, common factors and multiples, and powers (including square roots). • Understand operations on fractions, decimals, percentages, and integers. • Use rates and ratios. • Know commonly used fraction, decimal, and percentage conversions. • Know and apply standard form, significant figures, rounding, and decimal place value. Measurement • Select and use appropriate metric units for length, area, volume and capacity, weight (mass), temperature, angle, and time, with awareness that measurements are approximate. • Convert between metric units, using decimals. • Deduce and use formulae to find the perimeters and areas of polygons and the volumes of prisms. • Find the perimeters and areas of circles and composite shapes and the volumes of prisms, including cylinders. Statistics Statistical investigation • Plan and conduct surveys and experiments using the statistical enquiry cycle: • determining appropriate variables and measures; – considering sources of variation; – gathering and cleaning data; – using multiple displays, and re-categorising data to find patterns, variations, relationships, and trends in multivariate data sets; • comparing sample distributions visually, using measures of centre, spread, and proportion; • presenting a report of findings. Statistical literacy • Evaluate statistical investigations or probability activities undertaken by others, including data collection methods, choice of measures, and validity of findings.

ISBN: 9780170217095

The Olympic Games

Do you know the answers to the following questions?

- What does 'Olympic' mean?
- When and where did the Olympic Games begin?
- How often are they held?
- Where are the next summer Olympic Games?
- Is there a winter Olympic Games?
- What is the difference between the summer and winter Olympic Games?
- Who participates in the Olympics?
- What is the Olympic symbol and what does it mean?
- What is the significance of the Olympic flame?

The Olympic Games are an exciting celebration of world sporting achievements and an opportunity for countries to show themselves to the world. The following exercises will give you some knowledge of the Games, answers to the above questions and more.

Exercise 1

Choose an aspect of the Olympic Games, research it and prepare a presentation. You can select one of the suggestions above or choose one of your own.

ISBN: 9780170217095

Past Olympic Games

- Where and when were the last Olympic Games held?
- Have the Games been held in New Zealand?
- Have the Games been held in London before 2012?

Below is a table of cities that have previously hosted the modern summer Olympic Games.

Year	City	Year	City	Year	City
1896	Athens	1936	Berlin	1980	Moscow
1900	Paris	1948	London	1984	Los Angeles
1904	St Louis	1952	Helsinki	1988	Seoul
1908	London	1956	Melbourne	1992	Barcelona
1912	Stockholm	1960	Rome	1996	Atlanta
1920	Antwerp	1964	Tokyo	2000	Sydney
1924	Paris	1968	Mexico City	2004	Athens
1928	Amsterdam	1972	Munich	2008	Beijing
1932	Los Angeles	1976	Montreal	2012	London

Exercise 2

1. Why are they called the modern summer Olympic Games?
2. Name the countries of each of these host cities.
3. How many Games have been celebrated in the southern hemisphere?
4. Can you give an explanation why most of the Games have been held in the northern hemisphere?
5. What percentage have been held in the United States? (answer to one decimal place)
6. How many cities have held the Olympic Games more than once (including the 2012 London Olympics)? What percentage is this?
7. Mark each of the cities in the table above on a map of the world.
8. Why were the Games not celebrated in 1916, 1940 or 1944?

ISBN: 9780170217095

Roman numerals

The Olympic Games are signified by a number written in Roman numerals, with Athens being the 1st (I) Olympiad and London the 30th (XXX) Olympiad. Although the 1916, 1940 and 1944 Games were not held, these dates are still recognised as Olympiad years.

The Roman numbering system uses the following symbols:

1	I
4	IV
5	V
9	IX

10	X
11	XI
40	XL
49	XLIX

50	L
100	C
500	D
1000	M

You will notice a system of subtraction and addition was used, and the order of the symbols is important. For example 69 = LXIX, 495 = VD. Use the tables above to answer the questions that follow.

Exercise 3

1 Why are Roman numerals used?

2 What are Roman numerals for the following?

a 6 **c** 33 **e** 99
b 20 **d** 60 **f** 450

3 What numbers do these Roman numerals represent?

a XIV **c** XCV **e** DCCVII
b XXXVIII **d** CMII **f** MM

4 Write the Olympiad of the following cities in Roman numerals:

a Atlanta **c** Amsterdam **e** Rome
b Melbourne **d** Moscow **f** Beijing

5 The year of the 1900 Paris Olympics written in Roman numerals is MCM. What is the year of the following Olympic Games written in Roman numerals?

a Antwerp **c** Mexico City **e** Stockholm
b London (1948) **d** Barcelona **f** Seoul

6 What is today's date (day/month/year) in Roman numerals?

7 What number Olympiad is London 2012?

8 Calculate the following and give your answer in Roman numerals:

a $4 + 3 \times 2$ **c** $72 - \sqrt{36}$ **e** LXI + LIV
b $5(6 - 3)$ **d** XIV + XLVII **f** CLX - XXVI

ISBN: 9780170217095

The London Olympics

London, England has been chosen to host the XXXth Olympiad. Its opening ceremony is to be held on 27 July and the closing ceremony on 12 August 2012. It is expected to be the biggest event ever held in London, with an estimated 17 000 athletes from 205 countries participating in 302 events in 26 sports. Of course it will attract many thousands of extra tourists before, during and after the Games, somewhere in the vicinity of two to three million visitors. The London Olympics will also create many thousands of extra employment opportunities, short and long term, some paid employment and some voluntary work.

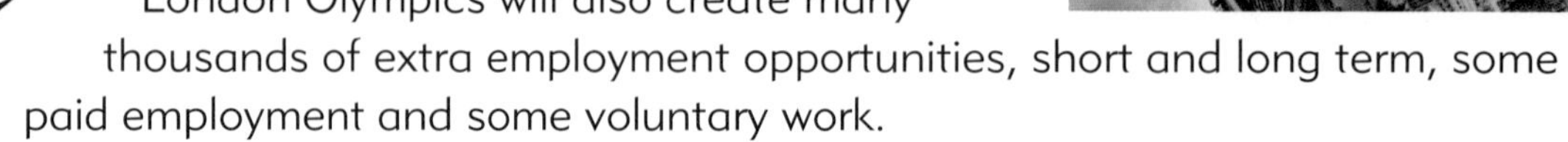

- Where is London?
- How far away is it from Auckland, Sydney, Moscow?
- Approximately how much does it cost to fly there?

Exercise 4

1 Using your copy of the world map you began on page 10, mark these places on your map.

a Sydney
b Auckland
c London
d New York
e Tokyo
f Paris
g Beijing
h Moscow
i Canada
j Cape Town
k Kenya
l Chile

ISBN: 9780170217095

2 Use the table below containing statistics about London to calculate answers to the questions that follow.

From London to …	Distance (km)	Time taken	Average adult airfare (one way, $NZ)
Auckland	18 363	28 hr 30 min	$3100
Sydney	17 018	24 hr 45 min	$2800
New York	5578	14 hr 25 min	$1083
Beijing	8153	13 hr 40 min	$1669
Paris	343	1 hr 15 min	$294
Cape Town	9684	11 hr 35 min	$1804
Moscow	2504	4 hr	$751
Los Angeles	11 681	19 hr 5 min	$2023
Santiago	8769	10 hr 5 min	$1274

LONDON

a Distance between London and Auckland?
b Distance between London and Moscow?
c Distance between London and Beijing?
d Time taken to fly from Cape Town to London?
e Time taken to fly from Auckland to London?
f Two adult fares from Paris to London?
g Three return adult fares from Moscow to London?
h Two return adult fares from New York to London?
i Children between the ages of two and 11 years pay 67% of the adult fare. What is the return fare of a New Zealand ten-year-old to the London Games?
j **i** The Morrison family (two adults, a 15-year-old and a ten-year-old) from Tauranga, New Zealand have reserved air tickets to London. What will the return fares cost?
ii They are offered a 15% discount if they pay in full, 12 months before departure. How much will their tickets cost now?
iii Travel insurance for 20 days is approximately 15% of the paid air fare price. Use your answer to ii and calculate how much the insurance will cost.
k It costs $3100 to travel 18 363 km from Auckland to London, which is approximately 17 cents per km. Calculate the cost per kilometre for these cities to London:
i Sydney **ii** Beijing **iii** Paris
l It takes four hours to travel the 2504 km from Moscow to London, which is an average speed of 626 km/h. Calculate the km/h for these travel distances to London: (answer to nearest whole number)
i Auckland
ii Santiago (round time to nearest hour)
iii Los Angeles (round time to nearest hour)
iv Can you give an explanation why there is a significant difference in speed?

ISBN: 9780170217095

Time differences from London

Study this world map and the time differences in relation to London, England. Time around the world is measured from Greenwich, London. This is called Greenwich Mean Time (GMT).

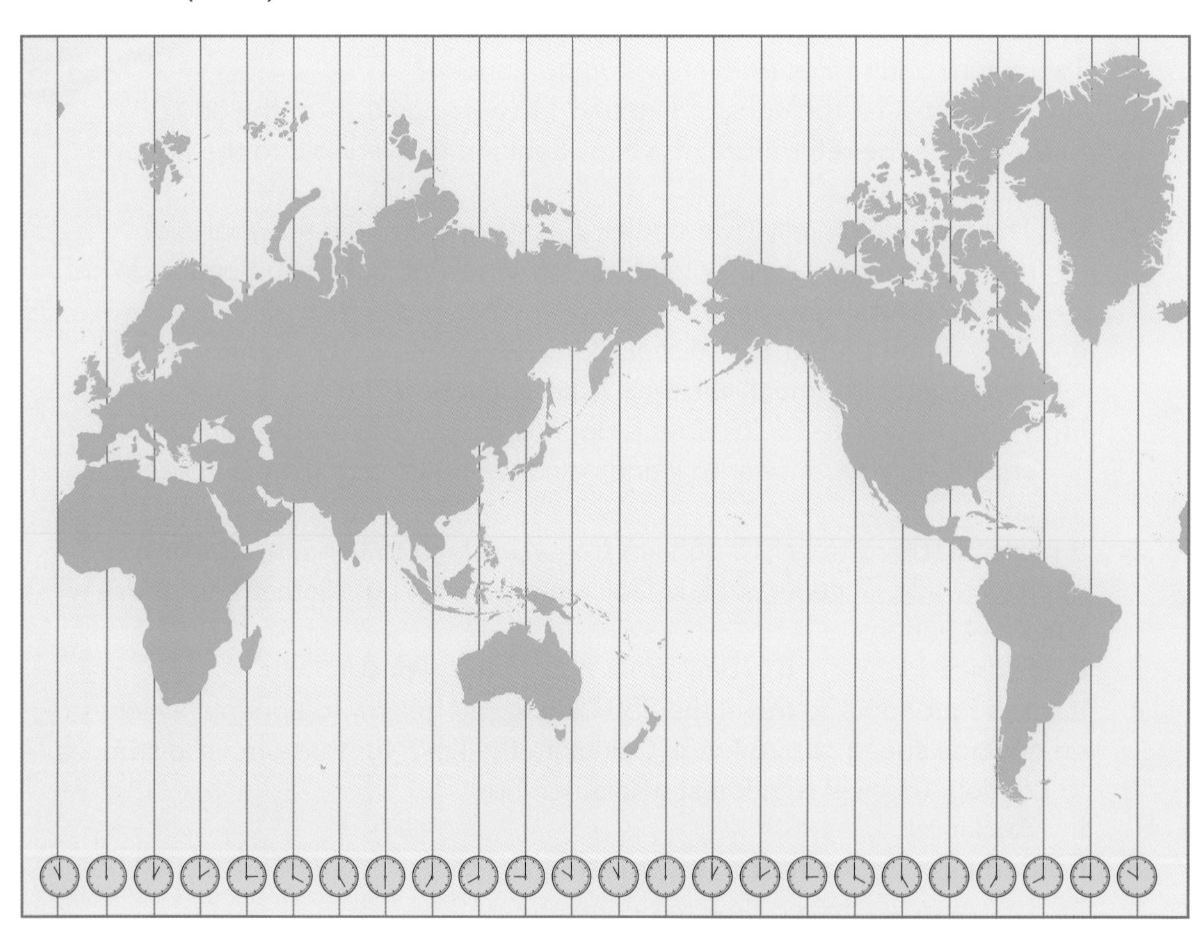

ISBN: 9780170217095

Exercise 5

1 What is the time difference between London and ...

a	Auckland?	**c**	Paris?	**e**	New York?
b	Sydney?	**d**	Beijing?	**f**	Moscow?

2 If it is 9 a.m. (0900 hrs) GMT in London, what is the time in ...

a	Auckland?	**c**	Paris?	**e**	New York?
b	Sydney?	**d**	Beijing?	**f**	Moscow?

3 If it is 7.30 p.m. (1930 hrs) in London on 27 July 2012, what is the time and date in Auckland?

4 If I want to watch the following programmes live (London GMT time) on television, what time and day would I need to tune in for each of the following countries:

a Swimming, live in London 10:00 on 30 July (in Sydney)
b Athletics, live in London 18:50 on 5 August (in Paris)
c Weightlifting, live in London 15:30 on 5 August (in Moscow)
d Triathlon, live in London 09:00 on 4 August (in Auckland)
e Gymnastics, live in London 13:30 on 12 August (in Tokyo)
f Equestrian, live in London 10:30 on 31 July (in Perth)
g Tennis, live in London 12:00 on 3 August (in New York)
h Rowing, live in London 09:30 on 4 August (in Berlin)
i Track cycling, live in London 16:00 on 7 August (in Los Angeles)
j Opening ceremony, live in London 19:30 on 27 July (in Cape Town)
k Closing ceremony, live in London 19:30 on 12 August (in Beijing).

ISBN: 9780170217095

Getting around London

The London Olympic Park is a new development in East London, Stratford City. The Olympic Stadium and Athlete's Village are located in the park, as well as nine of London's 13 sporting venues. Nine others outside London will host sports, in the City of Coventry, Eton Dorney, Hadleigh Farm, Hampden Park (Scotland), St James' Park, Millennium Stadium (Wales), Lee Valley White Water Centre, Weymouth and Portland and Old Trafford.

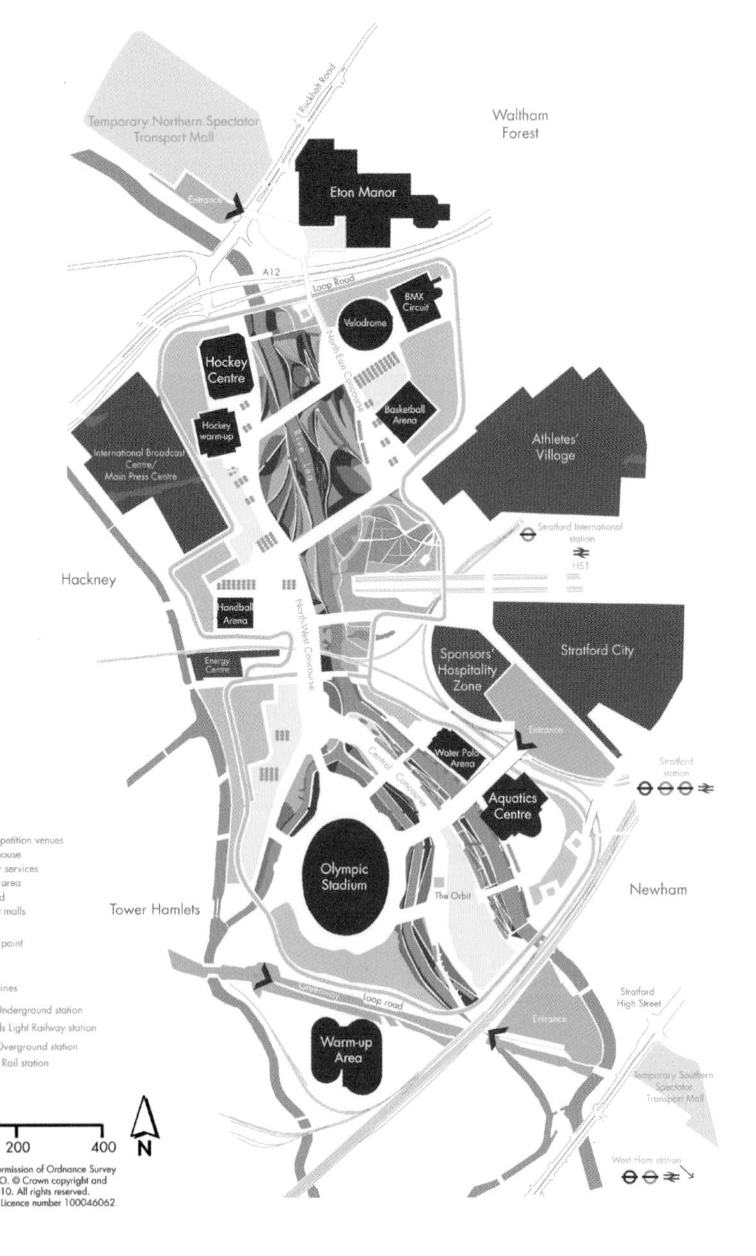

ISBN: 9780170217095

Exercise 6

1. List the sports being hosted in the Olympic Park.
2. What sports are being held at ...
   a Old Trafford?
   b Hyde Park?
   c The Royal Artillery Barracks?
   d Horse Guards Parade?
3. Estimate the distance from the Athlete's Village to the Olympic Stadium.
4. Estimate the distance from Stratford Street Station to the Olympic Stadium.
5. Estimate the area of the Olympic Park.
6. Below is a map of the London venues.

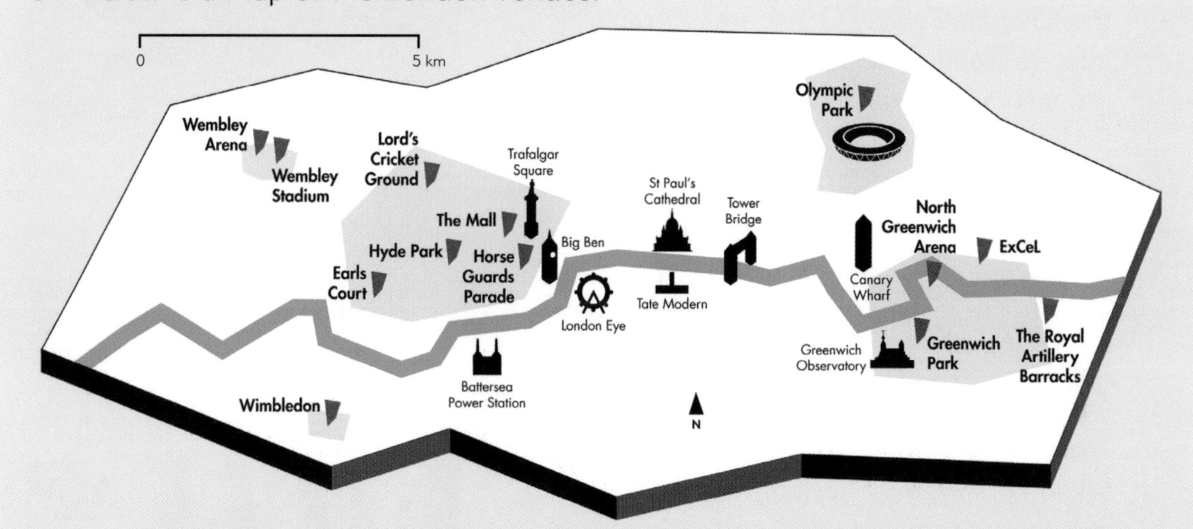

   a Olympic Park is northeast of St Paul's Cathedral. What direction are these venues from St Paul's Cathedral?
      i Canary Wharf
      ii Lord's Cricket Ground
      iii Wimbledon
      iv Hyde Park
      v Royal Artillery Barracks.
   b What do these places have in common: St Paul's Cathedral, Greenwich Observatory, Canary Wharf, London Eye, Trafalgar Square and Battersea Power Station?
7. The Morrison family have four tickets to each of the following events.

Sport and venue	Date and time	Price per ticket
Swimming	30 July 19:30	£95
Athletics	3 August 10:00	£65
Rowing	4 August 09:30	£50
Athletics	6 August 18:50	£150
Weightlifting	7 August 15:30	£30
Athletics	10 August 19:00	£125
Closing ceremony	12 August 19:30	£150

   a What is the total cost in £?
   b If £1 is equivalent to NZ$2.02617, what is the total cost of their tickets in NZ$?

ISBN: 9780170217095

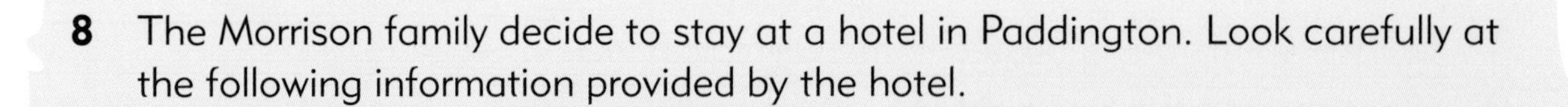

8 The Morrison family decide to stay at a hotel in Paddington. Look carefully at the following information provided by the hotel.

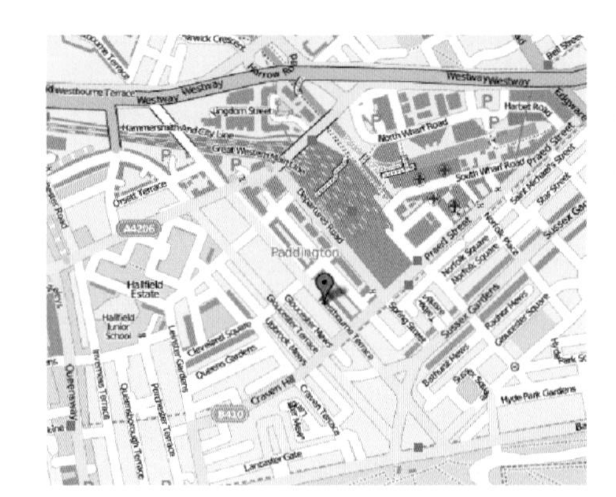

Hyde Park Premier London Paddington
14–16 Craven Hill, Lancaster Gate,
London W2 3DU
Situated in the Bayswater-Paddington neighbourhood, this hotel is close to Kensington Gardens, Royal Albert Hall and Buckingham Palace. Other area attractions include Trafalgar Square and Big Ben. The property is also just a short walk from Paddington main line, with direct links to Heathrow Airport terminal and several underground stations.

Facilities

- Voicemail
- Safe
- Data ports
- Wakeup service
- Electronic room keys
- Inhouse bar
- Clothing iron
- Mini-bar in room
- Coffee/tea maker
- TV in room
- Hair dryer
- Room service

Hotel bar/lounge
Hyde Park Premier London Paddington provides a concierge desk, a bar/lounge, a coffee shop and laundry facilities.

LCD television
LCD televisions include premium satellite channels. Guest rooms are equipped with premium bedding, and bathrooms with shower/tub combinations and makeup/shaving mirrors.

Nearby points of interest

- Kensington Gardens (0.7 km)
- Hyde Park (1.1 km)
- Royal Albert Hall (1.3 km)
- Marble Arch (1.5 km)
- Lord's Cricket Ground (1.9 km)
- Regent's Park (2.1 km)
- Kensington Palace (1 km)
- Albert Memorial (1.3 km)
- Hyde Park Speakers' Corner (1.4 km)
- Portobello Road Market (1.7 km)
- Harrod's (1.9 km)
- Madame Tussaud's Wax Museum (2.1 km)

Nearest major airports

- London (LCY-London City): 16 km
- London (LHR-Heathrow): 19.2 km
- London (LGW-Gatwick): 39.7 km

Room type	Rate per room with two guests
Deluxe double room	NZ$291 ($262 without breakfast)
Deluxe single room (twin-bed)	NZ$256 ($231 without breakfast)

ISBN: 9780170217095

a What is the address of the Hyde Park Premier London Paddington?

b The Morrison family — two adults, 15-year-old and ten-year-old — decide to stay for 20 days from 26 July to 13 August 2012, in one double and one single room (without breakfast). How much will their accommodation cost them?

c Name five complimentary facilities available.

d Name five tourist attractions less than 2 km from the hotel.

e Which London airport is closest to the Hyde Park Premier London Paddington?

f The Morrison family plan to travel from the hotel to Stratford Park, which is the preferred station for the Olympic Park. Using public transport, research the travelling options available and which option could be best.

ISBN: 9780170217095

The Olympic torch and oath

- What is the significance of the Olympic torch?
- Name some famous athletes who have carried the torch?
- Where does it begin?
- How long does it burn for?

Exercise 7

1. Research the path the Olympic flame will take for the London Games. Draw this path onto a world map. How long will it take? How many runners are involved in the relay?
2. Suggest several ways the torch is kept alight.
3. Write 2–4 sentences about the significance of the Olympic oath.
4. What is the Olympic motto in both English and Latin?

The Olympic Oath

'In the name of all competitors, I promise that we will take part in these Olympic Games, respecting and abiding by the rules which govern them, in the true spirit of sportsmanship, for the glory of sport and the honour of our teams.'

ISBN: 9780170217095

Olympic symbols

During the London Games many participants and visitors will visit other parts of the United Kingdom. Many of these people will not have English as their first language, and therefore will need to find their way around using internationally recognised symbols.

Exercise 8

Study the symbols above and select the appropriate one for each of the following situations.

1. You need to buy a ticket to an Olympic event.
2. You need to email home.
3. You have lost your suitcase.
4. You need to have a snack.
5. You need first aid.
6. You need a taxi.
7. You need the police.
8. You need the information centre.
9. You need a train.
10. You want to find an elevator (lift).

ISBN: 9780170217095

11 Design your own logo for any of the above.

12 Use the diagram below to answer the questions that follow.

- **a** What does this symbol mean?
- **b** Construct it using the correct colour combination.
- **c** How many lines of symmetry does it have?
- **d** Copy the following and reflect in the:
  - **i** x-axis
  - **ii** y-axis.

ISBN: 9780170217095

e **i** Copy and complete this table showing the number of intersections and circles if the symbol was continued to be drawn.

Circles (C)	Intersection (I)
2	2
3	4
4	6
5	
6	
7	
8	
n	

ii Are you able to write a rule linking circles and intersections together?

f **i** Copy the Olympic symbol into your book and colour the nine regions so that no two colours touch along borders. You must use the least number of colours possible.

ii If there were only three circles, what is the least number of colours needed?

iii If there were seven circles, what is the least number of colours needed?

iv Copy this table into your book and complete.

Circles (C)	Least number of colours needed
1	1
3	2
5	
7	
9	
n	

ISBN: 9780170217095

Participating countries

In 2011 there were 205 countries registered with the International Olympic Committee. Most of these countries will hope to send athletes to the 2012 London Olympic Games. At this stage an estimated 17 000 athletes are expected to attend the Games.

Registered Countries for the 2012 London Olympic Games

Afghanistan
Albania
Algeria
American Samoa
Andorra
Angola
Antigua and Barbuda
Argentina
Armenia
Aruba
Australia
Austria
Azerbaijan
Bahamas
Bahrain
Bangladesh
Barbados
Belarus
Belgium
Belize
Benin
Bermuda
Bhutan
Boliva
Bosnia & Hercegovina
Botswanna
Brazil
British Virgin Islands
Brunei
Bulgaria
Burkin Faso
Burundi
Cambodia
Cameroon
Canada
Cape Verdi
Cayman Islands
Chad
Chile
China
Chinese Taipei
Colombia
Comoros
Congo – Dem Republic
Cook Islands
Costa Rica
Croatia
Cuba
Cyprus
Czech Republic
Democratic People's Republic of Korea
Denmark
Djibouti
Dominica
Dominican Republic
Ecuador
Egypt
El Salvador
Equatorial Guinea
Eritrea
Estonia
Ethiopia
Fed. States of Micronesia
Fiji
Finland
France
Gabon
Gambia
Georgia
Germany
Ghana
Great Britain
Greece
Grenada
Guam
Guatemala
Guinea
Guinea-Bissau
Guyana
Haiti
Honduras
Hong Kong
Hungary
Iceland
India
Indonesia
Iran
Iraq
Ireland
Israel
Italy
Ivory Coast
Jamaica
Japan
Jordan
Kazakhstan
Kenya
Kiribati
Korea
Kuwait
Kyrgyzstan
Lao People's Republic
Latvia
Lesotho
Liberia
Libyan Arab Jamahiriya
Liechtenstein
Lithuania
Luxembourg
Macedonia – Republic of
Madagascar
Malawi
Malaysia
Maldives
Mali
Malta
Marshall Ilands
Mauritania
Mauritius
Mexico
Monaco
Mongolia
Montenegro
Morocco
Mozambique
Myanmar
Nambia
Nauru
Nepal
Netherlands
Netherlands Antilles
New Zealand
Nicaragua
Niger
Nigeria
Norway
Oman
Pakistan
Palau
Palestine
Panama
Papua New Guinea
Paraguay
Peru
Philippines
Poland
Portugal
Puerto Rico
Qatar
Republic of Moldava
Romania
Russian Federation
Rwanda
Saint Kitts and Nevis
Saint Lucia
Saint Vincent and the Grenadines
San Marino
Sao Tome and Principe
Saudi Arabia
Senegal
Serbia
Seychelles
Sierra Leone
Singapore
Slovakia
Slovenia
Solomon Islands
Somalia
Spain
Sri Lanka
Sudan
Surinam
Swaziland
Sweden
Switzerland
Syria
Tajikistan
Tanzania
Thailand
Timor-Leste
Togo
Tonga
Trinidad & Tobago
Tunisia
Turkey
Turkmenistan
Tuvala
Uganda
Ukraine
United Arab Emirates
United States of America
Uruguay
Uzbekistan
Vanuatu
Venezuela
Vietnam
Virgin Islands
Virgin Islands
Yemen
Zambia
Zimbabwe

ISBN: 9780170217095

Exercise 9

1 Which country has the largest population?

2 If it is winter (July) in New Zealand, name two other countries where it will also be winter.

3 Name two countries participating in the Games where it will be summer in July.

4 If there are 17 000 athletes, 5000 team officials and 20 000 media expected at the Games, illustrate the number of athletes, officials and media on a pie graph.

5 If the total number of official participants is expected to be 42 000, what percentage of them would be athletes? (answer to one decimal place)

6 On the next page are the flags of some participating countries.

- **a** How many have only one line of symmetry?
- **b** How can you tell that some countries belong to the Commonwealth?
- **c** Construct the Union Jack (flag of the United Kingdom).
- **d** What do the parts of the New Zealand flag represent?
- **e** Construct and correctly colour the New Zealand flag.
- **f** Copy this flag. Reflect it first in the x axis, then in the y axis.

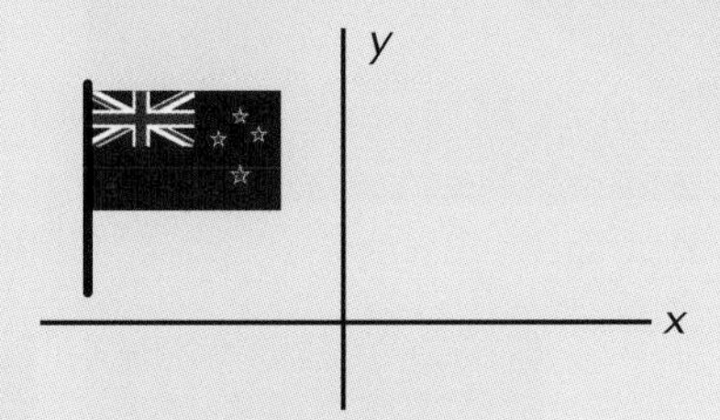

- **g** Position the flag so that point A is (1, 1).

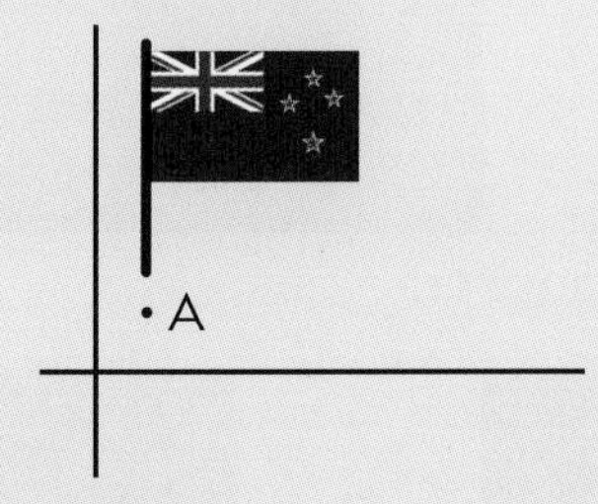

  - **i** Rotate it 90° anticlockwise about (0, 0).
  - **ii** Rotate it 180° about (1, 1).
- **h** Copy this flag.
  - **i** Enlarge it by scale factor 2.
  - **ii** Enlarge it by scale factor ½.

ISBN: 9780170217095

Flags of the world

Australia

Canada

China

Domincan Republic

Fiji

Finland

Greece

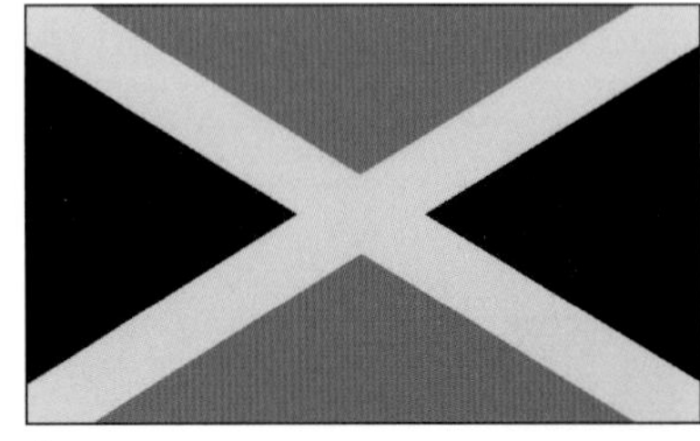
Jamaica

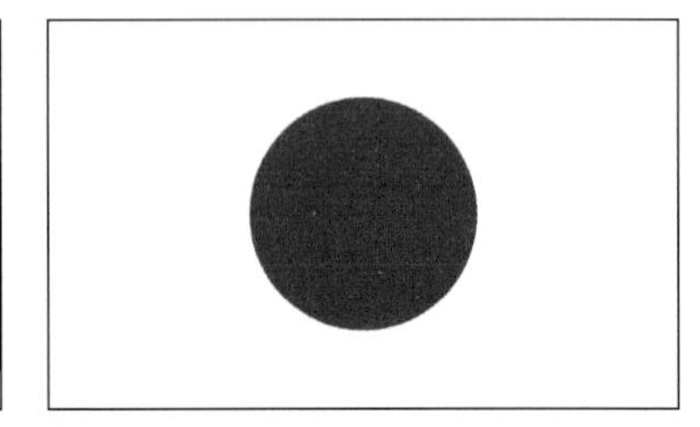
Japan

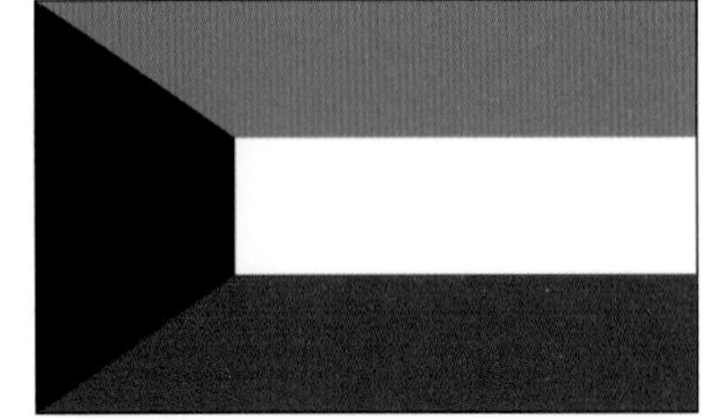
Kuwait

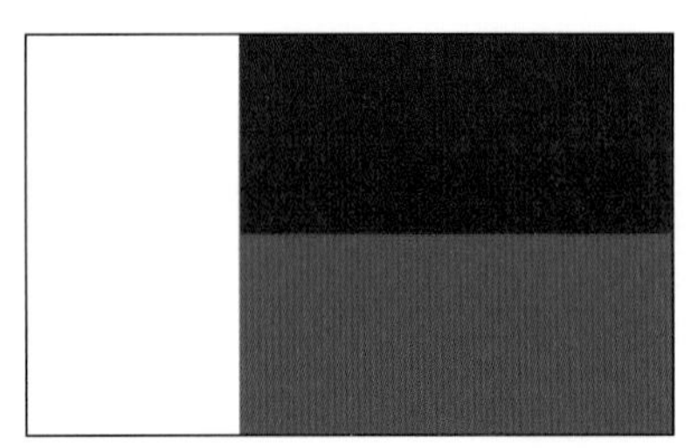
Madagascar

Mexico

New Zealand

Peru

Philippines

Poland

South Korea

United Kingdom

ISBN: 9780170217095

London 2012

The London 2012 Games will offer athletes the opportunity to compete in 26 different sports. The table below shows each discipline and the approximate number of competitors (M = men, W = women).

Archery: 128 (64M 64W)	**Fencing:** 212	**Swimming:** 950
Athletics: 2000	**Football:** 504 (288M 216W)	**Synchronised swimming:** 104W
Badminton: 172	**Gymnastics:** Artistic 196 (98M 98W) Rhythmic 96W Trampoline 32	**Table tennis:** 172 (86M 86W)
Basketball: 288 (144M 144W)	**Handball:** 336 (168M 168W)	**Taekwondo:** 128 (64M 64W)
Beach volleyball: 96 (48M 48W)	**Hockey:** 384 (192M 192W)	**Tennis:** 172 (86M 86W)
Boxing: 286 (250M 36W)	**Judo:** 386	**Triathlon:** 110 (55M 55W)
Canoe: Slalom 82 Sprint 246	**Modern pentathlon:** 72 (36M 36W)	**Volleyball:** 288 (144M 144W)
Cycling: BMX 48 Mountain bike 80 (50M 30W) Road 212 (145M 67W) Track 188 (104M 84W)	**Rowing:** 550 (353M 197W)	**Water polo:** 260 (156M 104W)
Diving: 136 (68M 68W)	**Sailing:** 380 (237M 143W)	**Weightlifting:** 260 (170M 90W)
Equestrian: Dressage 50 Eventing 75 Jumping 75	**Shooting:** 390	**Wrestling:** 344

ISBN: 9780170217095

Exercise 10

1. Which sport has the most athletes participating?
2. Which sport has the least athletes participating?
3. What percentage of the sports are women only? (answer to one decimal place)
4. Of the 17 000 expected athletes, what fraction are participating in athletics? (simplest form)
5. How many cyclists (men and women, all cycling disciplines) are expected to participate?
6. Design an appropriate logo for your town if it were to host the Olympics.
7. Every host city of the Olympic Games has a mascot. London has chosen Wenlock and Mandeville.
   - **a** What do these mascots signify?
   - **b** What would be an appropriate mascot if New Zealand were to host the Games?
8. The Olympic Stadium located at Olympic Park will be the venue for the opening and closing ceremonies, as well as the athletics and football finals. It will seat 82 000 and the expected average price of opening/closing ceremony tickets is £1000 each.
   - **a** If \$1 New Zealand is equivalent to 43 pence, how many New Zealand dollars is £1000 equivalent to?
   - **b** What will be the gross takings in New Zealand dollars from the opening/closing ceremonies, assuming all tickets are sold?
   - **c** 65 000 seats in the stadium are under cover. What percentage is this? (answer to nearest whole number)
   - **d** A 105 m x 68 m grass football field will be inside the stadium.
     - **i** What is its area?
     - **ii** What is its perimeter?
   - **e** The Olympic Stadium is expected to cost £690 million to build.
     - **i** Write this number in digits.
     - **ii** Write this number in standard form.
9. There is expected to be nine million tickets available for spectators to the various sporting events.
   - **a** Write this number in digits.
   - **b** If an average ticket costs £150, what is the total income from the ticket sales?
10. It is expected to cost UK£11.3 billion to build venues and host the Olympic Games. Write this number in standard form.
11. Below is a table showing the average temperatures and rainfall for London. The 2012 Olympic Games will be held in July.

Average rainfall for London

	Jan	Feb	Mar	Apr	May	Jun	Jul	Aug	Sep	Oct	Nov	Dec
Max (°C)	6	7	10	13	18	20	23	20	19	14	9	7
Min (°C)	2	3	4	6	8	12	14	12	10	8	5	4
Rain (mm)	50	42	37	38	42	50	60	58	48	58	62	45

ISBN: 9780170217095

a What will be the maximum average temperature?
b How much rain is expected?
c What is the range of temperatures in July–August?
d **i** Draw a bar graph to illustrate the maximum and minimum temperatures for the 12 months of the year.
ii Write 2–4 sentences about London's temperatures.

12 It is expected the Games will have a worldwide television audience of approximately four billion people. New Zealand has a population of 4.2 million.

a What fraction of this worldwide audience is New Zealand's population? (simplest form)
b The UK has a population of nearly 62 million. What fraction is this of the worldwide television audience? (simplest form)

13 Below is a table showing a timetable of some of the sporting events.

	July				**August**										
	28	29	30	31	1	2	3	4	5	6	7	8	9	10	11
Athletics							•	•	•	•	•	•	•	•	•
Swimming	•	•	•	•	•	•	•	•	•	•	•	•	•	•	
Canoeing		•	•	•	•	•	•	•	•	•	•	•	•	•	•
Equestrian	•	•	•	•	•	•	•	•	•	•	•	•	•		
Gymnastics	•	•	•	•	•	•	•	•	•	•	•	•	•	•	•
Rowing	•	•	•	•	•	•	•	•							
Cycling	•	•			•							•	•	•	•
Triathlon								•			•				
Weightlifting	•	•	•	•	•	•	•	•	•	•	•				
Tennis	•	•	•	•	•	•	•	•	•						
Volleyball	•	•	•	•	•	•	•	•	•	•	•	•	•	•	•
Shooting	•	•	•	•	•	•	•	•	•	•					

a Which sport has the longest competition?
b Which sport is completed first?
c How many sports are played on 1 August?
d What sport begins eight days after the opening ceremony?
e Name a day when there would be less than six sports taking place.
f Draw up a timetable so that each sport is seen and each day from 28 July to 11 August is occupied. Only one sport a day can be seen.

ISBN: 9780170217095

Athletics

On the track

Did you know ...

- Ray Ewry (USA) won ten gold medals in athletics between 1900 and 1908.
- Peter Snell won three gold medals for New Zealand in 1960 and 1964.
- Harry Kerr won New Zealand's first Olympic medal (bronze) in the 1908 London Games for the 3500 m walk.
- Nick Willis won a silver medal for New Zealand in the1500m at the 2008 Beijing Olympic Games.

Olympic Track Events	
100 metres	400 metre hurdles
100 metre hurdles	3000 metre steeplechase
110 metre hurdles	4 x 100 metre relay
200 metres	4 x 400 metre relay
400 metres	10 km walk
800 metres	20 km walk
1500 metres	50 km walk
5000 metres	42 km marathon
10 000 metres	

New Zealand athletes have traditionally been better at longer distances such as 1500 m, 10 000 m and the marathon.

- How many medals has New Zealand won on the track?
- Who were these athletes?
- Which of the above events do both men and women compete in?
- Who holds the men's 100 m Olympic record?
- Who holds the women's 100 m Olympic record?

ISBN: 9780170217095

Exercise 11

1. If one kilometre is 1000 m, how many events are longer than one kilometre?
2. If an athlete competed in a heat, semifinal and final of both the 100 m and 200 m, how far in kilometres would the athlete have run?
3. What is the total distance in kilometres run by the 4 x 400 m relay team?
4. Usain Bolt can run the 100 m in 9.69 seconds (sec).
   - **a** What is this time to the nearest tenth of a second?
   - **b** Using the answer to a, what is his speed in:
     - **i** m/s?
     - **ii** km/h?
5. Usain Bolt can run the 200 m in 19.30 sec. Is he faster over 100 or 200 m? What is the time difference?
6. An athletic track has a circumference of 400 m. How many laps would these races be?
   - **a** 1500 m
   - **b** 5000 m
7. If Usain Bolt takes 28 strides to complete the 100 m, how long is each stride in centimetres? (answer to nearest centimetre)
8. At the 2008 Beijing Olympics, Shelly-Ann Fraser of Jamaica won the womens 100 m in 10.7 sec, and Usain Bolt won the mens in 09.69 sec. What was the time difference?
9. Jamaica was very successful at the Beijing Olympics in the track events. The first two placings of the women's 100 m were Shelly-Ann Fraser (10.78 sec) and Sherone Simpson (10.98 sec). What is the time difference?
10. Jamaican athletes also dominated the 200 m, with Veronica Campbell-Brown winning gold (21.70 sec) and Kerron Stewart winning bronze (21.99 sec). What is the time difference?
11. The Jamaican men's 4 x 100 m relay team won a gold medal at the Beijing Olympics in a time of 37.10 sec, which is a world record. What is the average time for each runner?
12. Nick Willis of New Zealand won a silver medal in the men's 1500 m at the Beijing Games. His time was 3:34.16 minutes (min). The winner's time was 3:33.11 min. What is the time difference?
13. Lorraine Moller of New Zealand won a bronze medal in the marathon at the Barcelona Games in a time of 2:33.59. Describe this time in words.
14. One of New Zealand's greatest track athletes, Peter Snell, won a gold medal at Rome in the 800 m in 1:46.3, and four years later in Tokyo again won gold in the 800 m in 1:45.1. How much quicker was he in Tokyo?
15. Peter Snell also won a gold medal at the 1964 Tokyo Games in the 1500 m with a time of 3:38.1. Edwin Flack of Australia won the same event at the 1896 Athens Games in 4:33.2.
    - **a** How much faster was Peter Snell?
    - **b** What is this difference as a percentage of the 1896 time?
16. At the 1956 Melbourne Games, Norman Reid from New Zealand won a gold medal for the 50 km walk in 4:30:42.8. The second placegetter was 2 min 18.2 sec behind. What was the time of the second placegetter?

ISBN: 9780170217095

17 Use the diagram of the hurdles below to answer the questions that follow.

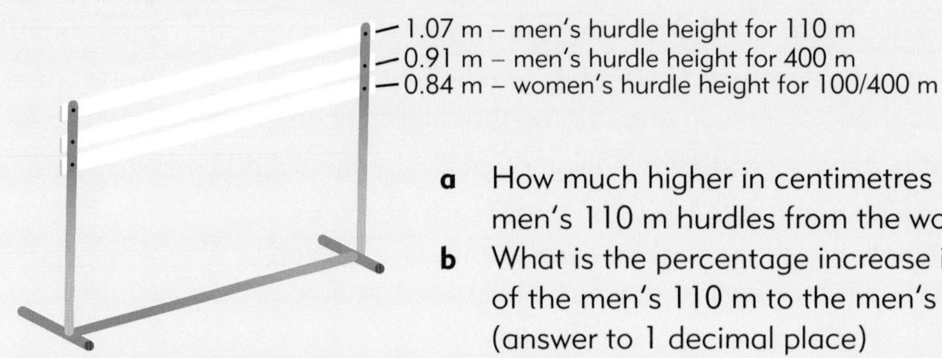

a How much higher in centimetres are the men's 110 m hurdles from the womens?

b What is the percentage increase in height of the men's 110 m to the men's 400 m? (answer to 1 decimal place)

18 In the women's 100 m hurdles each athlete runs 13 m to the first hurdle, then each of the ten hurdles are 8.5 m apart. Using a scale of 1 cm = 10 m, draw a scale of the 100 m hurdle race and calculate the length of the run to the finish line.

On the field

Did you know ...

- The triple jump was first introduced for women in 1996.
- The Fosbury Flop technique for high jump was developed by Dick Fosbury in 1968.
- Yvette Williams won New Zealand's first female gold medal in athletic field events for the long jump at the 1952 Helsinki Games.
- New Zealander Valerie Adams is the current World, Olympic and Commonwealth record holder for the women's shot put with a distance of 21.07 m.
- Beatrice Faumuina has represented New Zealand in three Olympics in discus and was a gold medallist in 1997 World championships.

Olympic Field Events	
Shot Put	Long Jump
Discus	High Jump
Hammer	Triple Jump
Javelin	Pole Vault

- Have New Zealand athletes won any medals in these field events?
- Which of these events do both men and women compete in?

Exercise 12

1 In 1896 the pole vault was won at a height of 3.30 m. In 1994 Sergey Bubka of the Ukraine set a World record of 6.14 m for the outdoor pole vault and 6.15 m for the indoor pole vault. What has been the percentage increase for the outdoor record from 1896 to 1994? (answer to the nearest whole number)

ISBN: 9780170217095

2 In 2000 the pole vault for women was introduced into the Games. The 2008 Beijing Games women's event was won with a World record height of 5.05 m. The men's pole vault was won with a height of 5.96 m, an Olympic record. What is the difference in centimetres?

3 Yvette Williams of New Zealand won a gold medal at the 1952 Helsinki Games for the long jump in a distance of 6.24 m. The distance at the 2008 Beijing Games was 7.04 m. From 1952 to 2008 there have been 14 Games. What has been the average increase in length over this time? (answer to one decimal place)

4 In the past ten Olympics Games the winning heights for the women's high jump were:

Year	Height (m)
2008	2.05
2004	2.06
2000	2.01
1996	2.05
1992	2.02
1988	2.03
1984	2.02
1980	1.97
1976	1.93
1972	1.92
1928	1.59

a Copy this table and convert the heights to centimetres.

b Draw a histogram to illustrate these heights in centimetres.

c Write 2–4 sentences about your graph.

d What is the percentage increase in height from 1928 to 2008? (answer to nearest whole number)

5 The men's shot put result table for six athletes was:

Athlete	Attempt 1	2	3	4	5	6	Best
1	20.16	20.84	20.40	20.96	20.91	20.89	
2	19.99	NT	20.15	20.18	20.35	20.48	
3	18.65	17.90	18.55	19.04	19.28	20.09	
4	20.98	21.70	NT	21.32	21.06	21.72	
5	21.03	21.45	21.91	21.85	21.96	21.09	
6	20.48	21.36	21.99	NT	21.84	21.97	

a What does NT mean?

b Calculate the best put for each athlete.

c Place them in order of first to sixth.

6 Below are the results for field events at the 2008 Beijing Games.

Field event	Women	Men
Discus	64.74	68.82
Shot put	20.56	21.51
Javelin	71.42	90.57
Hammer	-	82.02

ISBN: 9780170217095

a Draw the continuum line (below) into your book and show these results.
b Write 2–4 sentences about the results.

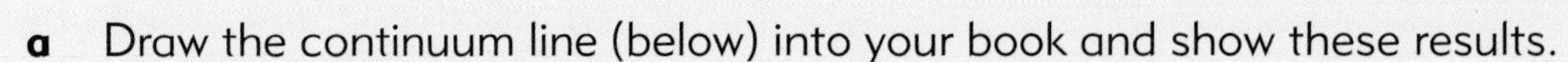

7 Al Oerter Jnr of the USA won four consecutive gold medals for the men's discus and each time set a new Olympic record. His distances were:

1956	1960	1964	1968
56.36 m	59.18 m	61.00 m	64.78 m

a At which Games did he make the biggest improvement?
b What was the difference in length between 1956 and 1968?
c Using the length in b, what was his average annual increase over the 12 year period?

8 The triple jump is made up of a hop, step and jump.
a When was it introduced into the Games for women?
b The three actions (hop, step, jump) are approximately in the ratios 1:3:4.
i How long would the hop be if the total length of the triple jump was 16.8 m?
ii If the length of the jump was 8.25 m, what is the length of the triple jump?

The marathon

Did you know ...

- The marathon is the oldest long distance running event.
- It was first run in 490 BC at a place in Athens named Marathon.
- The distance to run is 42.295 km. It was originally 40 km, but in 1908 the race was extended slightly so that British Royalty could watch the start from Windsor Castle.

New Zealand has won three bronze medals in the marathon:

- Barry Magee (1960, Rome)
- Mike Ryan (1968, Mexico City)
- Lorraine Moller (1992, Barcelona)

ISBN: 9780170217095

Exercise 13

1 At the 2008 Beijing Games the women's marathon was won by Constantina Dita of Romania in 2 hr 26.44 min, and the men's event was won by Samuel Kamau Wanjiru of Kenya in 2 hr 6.32 min, which was an Olympic record.

a How much faster was Samuel?

b Round Samuel's time to 2:07 and calculate his speed over 42 km in km/hr.

c Round Constantina's time to the nearest minute.

i What was her finishing time in minutes?

ii What was her average time to cover one kilometre? (answer in minutes and seconds)

2 The winner of the 1960 Rome Olympic marathon was Abebe Bikila. He ran barefoot in a time of 2 hr 15.16 min. At Tokyo in 1964 he repeated his victory in a World record time of 2 hr 12.11 min.

a How much faster did Bikila finish the race in 1964 compared to 1960?

b Write his 1964 time in words.

3 London's marathon map will look something like this.

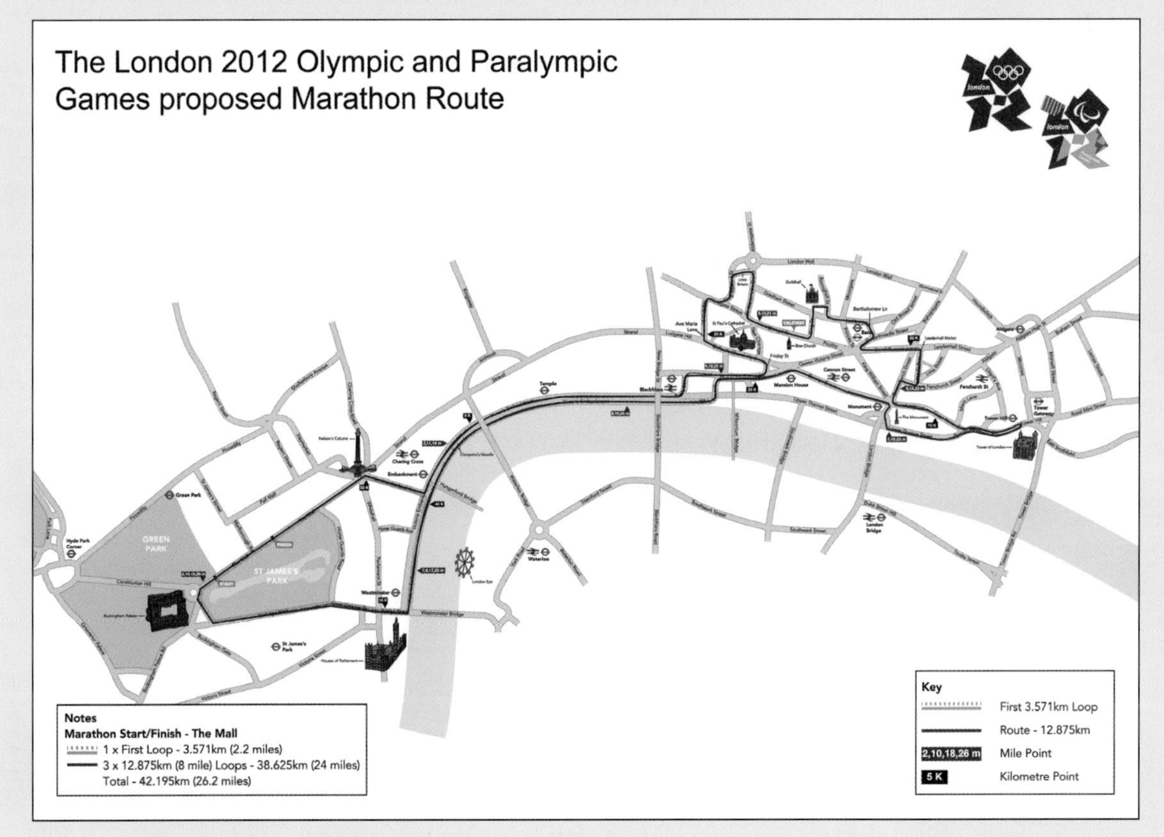

a If it takes a runner 50 minutes to reach the 10 km mark, what would the finishing time be?

b If it takes a runner two hours 45 minutes to complete the course, approximately what would be the times at the ...

i 10 km mark?

ii 20 km mark?

iii 30 km mark?

ISBN: 9780170217095

- **c** If a runner can run one kilometre in 4.5 minutes on average, what time would the runner expect to finish the marathon in?
- **d** At the 1992 Barcelona Games Lorraine Moller reached the 20 km mark in one hour 15 minutes. Her finishing time for the bronze medal was 2:33.59. What time did she run the final 22 km?
- **e** Where does the London 2012 marathon event start and finish?
- **f** What monument is at the 30 km mark?
- **g** How many 12.875 circuits of the course do runners complete?
- **h** How long is the marathon in miles?
- **i** The Olympic record for the men's marathon is 2:6.32 set in 2008. The women's record of 2:23.14 was set in 2000. What is the time difference?

The decathlon and heptathlon

Did you know ...

- The first multiple event of this kind – the pentathlon – was held by the ancient Greeks in 708 BC.
- The decathlon was introduced to the modern Games in 1912.
- The heptathlon was introduced for women in 1984.

The winners of these contests are often described as the greatest athletes in the world. The men's decathlon is a ten event contest, and the women's heptathlon is a seven event contest. Both competitions run over two days and points are awarded for the result achieved in each event. The athlete with the greatest number of points at the end of the contest is the winner.

Exercise 14

1 The points earned by gold and silver medallists for the decathlon at the 2008 Beijing Games are listed below. Use the table to answer the questions that follow.

	100 m	400 m	1500 m	Javelin	Pole vault	Discus	Long jump	High jump	Shot put	110 m hurdles
Bryan Clay (USA)	989	865	522	904	910	950	1005	794	868	984
Andrei Krauchanka (BIZ)	870	943	761	741	910	758	962	906	752	948

- **a** What were the total points for each athlete?
- **b** Which events did Bryan do better than Andrei?
- **c** Which event did they get equal points for?

ISBN: 9780170217095

2 The world champion decathlon athlete is Roman Sebrle of Czechoslovakia (TCH). His total points for the ten event contest is 9026. Have any other decathlon athletes passed the 9000 point mark? If yes, who?

3 If an athlete hopes to get 9000 points for the ten events, what must they average for each event?

4 One of the Britain's more famous athletes is Daley Thompson. In 1980 he won a gold medal for the decathlon and his points for the ten events were 8275.

a What did he average per event?

b Considering your answer to 1a above, how many more points would he need to get to win a gold medal at the 2012 London Games? (estimate to nearest whole number)

5 The points earned by gold and silver medallists for the heptathlon at the 2008 Beijing Games are listed below. Use the table to answer the questions that follow.

	100 m hurdles	200 m	800 m	Javelin	High jump	Long jump	Shot put
Natalie Dodrynska (UKR)	1059	944	855	833	978	1049	1015
Hyleas Fountain (USA)	1158	1058	886	704	1093	969	751

a What are the total points for each athlete?

b Which events did Hyleas do better than Natalie?

c Which event(s) does Hyleas need to improve?

6 One of the greatest women athletes of recent time is Jackie Joyner-Kersee (USA). Her Olympic and World record for the heptathlon of 7291 points, set at the 1988 Seoul Games, is still unbeaten.

a What did she average per event?

b Considering your answer to 5a above, how many more points would each athlete need to get per event if they were to break this Olympic record?

ISBN: 9780170217095

Swimming

Did you know ...

- Danyon Loader has achieved New Zealand's greatest swimming achievements in recent times. He won two gold medals at the 1996 Atlanta Olympic Games.
- At the 2008 Games in Beijing, Michael Phelps of the United States won an unprecedented eight gold medals.
- Natalie Couglin became the first United States female athlete to win six medals, and the first woman to win the 100 m backstroke gold in two consecutive Olympics.
- The oldest World record was broken in 2008 at Beijing by Great Britain's Rebecca Adlington. Her time of 8.14.10 for the 800 m freestyle beat the 1989 record by more than two seconds.

Olympic Swimming Events	
Freestyle	50 m, 100 m, 200 m, 400 m, 800 m, 1500 m
Backstroke	100 m, 200 m
Breaststroke	100 m, 200 m
Butterfly	100 m, 200 m
Medley	200 m, 400 m
Relay	4 x 100 m, 4 x 200 m
Medley Relay	4 x 100 m

ISBN: 9780170217095

Exercise 15

- How long is an Olympic pool?
- Can you name the four strokes in the medley?
- What order are they swum in?

1 How many lengths of the pool would you swim in the ...
  a 200 m breaststroke? **b** 1500 m freestyle?

2 **a** How many lengths of the pool would the team swim in the 4 x 200 m relay?
  b What is this distance in kilometres?

3 The winner's time for the men's 50 m freestyle at the 2008 Beijing Olympics was 21.30 sec, an Olympic record. The silver medallist's time was 21.45 sec. What was the time difference?

4 The following table shows the women's gold medal times at the 2000 Sydney, 2004 Athens and 2008 Beijing Olympic Games. Use the table to answer the questions that follow.

	2000	**2004**	**2008**
100 m freestyle	53.83	53.84	53.12 (OR)
100 m breaststroke	1.07.05	1:06.64	1:05.17 (OR)
100 m backstroke	1:00.21 (OR)	1:00.37	58.96
100 m butterfly	56.61 (WR)	57.72	56.73

  a Which stroke is the fastest?
  b Write the time for the 100 m backstroke for 2008 in words.
  c Which stroke had the biggest range in time over the three Games?
  d Which stroke had the smallest range in time over the three Games?
  e Write the time difference in words between the fastest and slowest in the 100 m breaststroke.
  f Which stroke was faster in 2000 than in 2008?

5 Michael Phelps' time for the 400 m medley at Beijing was 4.03.84, which is an Olympic and World record.
  a What is this time in minutes and seconds?
  b How many lengths of the pool is 400 m?
  c What is his average time per length of the pool?

6 The Netherlands women's 4 x 100 m freestyle relay team won at the 2008 Beijing Games in a time of 3 min 33.76 sec, an Olympic record time.
  a What was the average time for each swimmer? (answer to the nearest hundredth of a second)
  b How does this average time compare with results for the 100 m freestyle provided in question 4?

7 Danyon Loader's time for the 200 m freestyle at the 1996 Atlanta Olympics was 1 min 47.63 sec. Michael Phelps' time at the 2008 Beijing Olympics for the 200 m freestyle was 1 min 42.96 sec. How much faster was Phelps than Loader?

ISBN: 9780170217095

Diving

Did you know ...

- Diving is a form of gymnastics. Swedish and German gymnasts practised their skills by diving into the sea.
- Diving became a sport in its own right just over 100 years ago.
- There are two separate events — springboard and platform or high board diving.

Degree of difficulty

Some dives are more difficult than others and will therefore give the competitor a higher score. An easy dive performed well, however, can score more points than a difficult dive performed less well.

Example of scoring for competitor A	
9.0, 8.5, 8.9, 8.8, 9.0, 9.1, 8.8 (discard the highest and the lowest scores)	
9.0 + 8.9 + 8.8 + 9.0 + 8.8	44.5
44.5 x 3 (degree of difficulty)	133.5
133.5 x 0.6	80.1
Final score	**80.1**

The judges award each dive a mark out of ten. The highest and lowest scores are removed, and the remaining scores are added together and multiplied by the degree of difficulty of the dive (a number up to 3.5). Finally this number is multiplied by 0.6 to get the final score.

ISBN: 9780170217095

Exercise 16

1 Use the method outlined above to calculate this diver's final score. (answer to two decimal places)

a 7.8, 7.9, 8.1, 7.9, 8.0, 8.1, 8.3 (Difficulty 2.5)
b 8.5, 8.7, 8.8, 8.8, 8.5, 9.2, 8.4 (Difficulty 2.8)
c 9.0, 9.1, 8.9, 9.0, 9.1, 9.2, 9.2 (Difficulty 3.1)
d 8.4, 8.3, 8.5, 8.6, 8.3, 8.9, 8.8 (Difficulty 3)

2 Copy the following table into your book and use it to answer the questions below.

Competitor	Degree of difficulty	Judges scores							Final score
1	2.6	8.6	8.5	8.6	8.7	8.3	8.4	8.5	
2	3.0	8.7	8.7	8.8	8.9	8.6	8.9	9.0	
3	2.9	9.1	9.0	9.2	9.0	8.9	9.3	9.1	
4	2.8	9.4	9.5	9.3	95	9.6	9.5	9.5	
5	3.1	8.9	8.9	8.8	9.0	8.9	9.2	9.1	
6	3.2	9.2	9.1	9.2	9.2	9.5	9.4	9.4	
7	2.9	9.0	9.2	9.3	9.2	9.2	9.3	9.4	
8	2.5	9.8	9.9	9.8	9.8	9.7	9.8	9.6	

a Calculate each competitor's final score. (answer to two decimal places)
b Put the competitors in order from first to eighth.

ISBN: 9780170217095

12 Gymnastics

Did you know ...

- Nadia Comaneci of Romania won three gold medals in 1976 and became the first gymnast to score a perfect 10 in an Olympic competition.
- In Ancient Greece, gymnastics was used as a way of training young men to be fit for battle.
- Larissa Latynina of the Soviet Union won 18 medals – including nine gold – in the 1956, 1960 and 1964 Games.
- Artistic gymnast Nastia Liukin of the USA won a gold, three silver and a bronze medal at the 2008 Beijing Olympics.

Men's Gymnastic Events	
Horizontal bar	Parallel bars
Horse vault	Rings
Pommel horse	Floor exercises

Women's Gymnastic Events	
Horse vault	Asymmetrical bars
Balance beam	Floor exercises
Rhythmic all round	

Competitors can win medals for individual events as well as accumulating points for an all round final and team events.

Exercise 17

1 Draw to scale the balance beam used by women. It is 5.0 m long, 1.24 m off the floor and only 10 cm wide! (Use 1 cm = 20 cm.)

2 The results from each of the ten judges for Competitor 4 at the horse vault were:

9.35	9.50	9.55	9.65	9.75	9.80	9.20	9.75	9.30	9.90

a Calculate the average score given to the competitor.

ISBN: 9780170217095

b What score would the competitor receive if these were her judges scores for the floor: 8.95, 8.80, 8.75, 8.85, 8.75, 8.80, 8.70, 9.5, 9.15, 8.95.

3 The top ten competitors in the men's pommel horse were given the following scores. Use these to write out the placings for the top five competitors.

Competitor	1	2	3	4	5	6	7	8	9	10
	9.52	9.80	9.18	9.85	9.9	9.65	9.75	9.8	9.95	9.85

4 These were the results of two teams in the women's team floor exercises competition.

Competitor	1	2	3	4
China	9.5	9.15	8.85	9.82
Romania	9.34	9.45	9.05	9.35

a Which was the better team score?
b What was the difference in the team totals?
c Which team was more consistent?
d What was the average score for the Romanian team?

5 The marking schedule in gymnastics has four components to make up a total of ten points.

- Difficulty up to 3.4 points
- Construction up to 1.6 points
- Execution up to 4.4 points
- Risk, originality up to 0.6 points

Two competitors were given these points by five judges.

Competitor 1					
Judge	Difficulty	Construction	Execution	Risk	Total
1	2.8	0.8	2.8	0.1	
2	2.9	0.9	2.8	0.2	
3	2.6	0.7	2.6	0.1	
4	2.5	0.8	2.6	0.1	
5	2.9	1.2	3.4	0.2	

Competitor 2					
Judge	Difficulty	Construction	Execution	Risk	Total
1	2.9	0.9	3.0	0.2	
2	3.1	1.0	3.3	0.4	
3	3.0	1.0	3.2	0.3	
4	2.8	0.5	2.6	0.1	
5	2.9	0.9	2.9	0.2	

a Total each judge's score for each competitor.
b What was the grand total for each competitor?
c Who was the better competitor?

ISBN: 9780170217095

13 Weightlifting

Did you know ...

- Top heavyweights lift above twice their own bodyweight.
- In the 1896 Olympics there were just two events: one-arm and two-arm lifts.
- There are now ten classes, determined by bodyweight.
- Each competitor competes in two events, The Snatch and The Clean and Jerk, and it is on the combined scores of these that the winner is decided.
- Each competitor has three attempts at each weight.
- If competitors have lifted the same total weight, the winner is the competitor with the lowest bodyweight.
- Weightlifting has been a sport for over 2000 years.

Exercise 18

1 If two competitors each lifted a total of 150 kg (kilogram) in the 83 kg class, but one competitor weighed 82.5 kg and the other 82 kg, who would win?

2 The following were the results of the super heavyweight (over 110 kg) for the top three competitors.

ISBN: 9780170217095

	The Snatch						The Clean and Jerk					
Weight class (kg)	162.5	165	167.5	170	172.5	175	227.5	230	232.5	235	237.5	240
1	✓	✓	✓	✓	✓	XXX	✓	✓	✓	X✓	XXX	
2	✓	X✓	✓	✓	✓	XXX	✓	✓	✓	✓	XX✓	XXX
3	✓	✓	XX✓	X✓	XXX		✓	✓	X✓	XX✓	XXX	

a Which competitor won The Snatch? Why?
b Which competitor won The Clean and Jerk?
c What were the total weights for each competitor?
d Place the competitors in first to third positions.
e How much less than a tonne was the winner's total lift?

3 The classes for men's weightlifting are up to these weights: 56 kg, 62 kg, 69 kg, 77 kg, 85 kg, 94 kg, 105 kg and over 105 kg.
a Draw a histogram to illustrate these classes of weights.
b Which classes have the greatest range of weights?
c If there are approximately 2.2 lb (pounds) to 1 kg (kilogram), what is 10 kg in pounds?
d The 2008 winner of the men's 56 kg class, Long Qingyuan, lifted 292 kg. What is the proportion of his lift to his weight?
e Each of the ten classes has a name, for example over 110 kg is the super heavyweight. What are the other names?

4 Today competitors lift weights with a barbell. What would they have used 2000 years ago?

5 The weights at each end of the barbell are made up of 50 kg, 20 kg, 10 kg, 5 kg, 1 kg and 0.5 kg discs. Remembering that the weights need to be split evenly at each end of the barbell (so for example 150 kg total must have 75 kg at each end), what weight discs would you choose for ...
a 175 kg?
b 210 kg?
c 189 kg?
d 97 kg?

6 Draw this continuous line into your book and put on it these weights: 170, 110.5, 185, 162.5, 125, 147.

100 kg — 150 kg — 200 kg

ISBN: 9780170217095

Equestrian

Did you know ...

- The first equestrian event in ancient Olympics was a four-horse chariot race in 680 BC.
- Equestrian events have three courses – show jumping, dressage and the three-day event – and has individual and team sections.
- Raimondo d'Inzeo of Italy participated in eight Olympic Games from 1948 to 1976.
- Lorna Johnstone of Great Britain is the oldest female competitor at the Games. She was 70 years old in 1972.
- New Zealand has won 17 medals in equestrian.
- In 1996 the New Zealand equestrian team finished with a number one world ranking in the teams event, and the top three individual rankings were all from New Zealand: Blyth Tait (1), Mark Todd (2) and Andrew Nicholson (3).

Exercise 19

1 Use the following New Zealand equestrian medal table to answer the questions that follow.

	Team			**Individual**		
	Gold	Silver	Bronze	Gold	Silver	Bronze
2000 Sydney						I
1996 Atlanta			IIII	I	I	
1992 Barcelona		III				I
1988 Seoul			IIII	I		
1984 Los Angeles				I		

ISBN: 9780170217095

a How many were gold?
b What percentage were gold?
c Has New Zealand been more successful in teams or individual events? Explain your answer.
d Which Games were the most successful for the New Zealand team? Explain your answer.
e At the 1996 Atlanta Olympics there were seven members in the team. One member was awarded two medals.
  i Do you know who this person was?
  ii How many of the team received medals?
  iii What percentage of the team received medals? (answer to one decimal place)
f Do you know who the gold medallist was in 1984 and 1988?

2 In the show jumping event, if a horse knocks down part of an obstacle the competitor collects four faults, a refusal collects three faults, a second refusal six faults and a third disqualification. If a horse puts a foot in the water at the water jump, four faults are awarded. Calculate the following faults awarded to these competitors:
a Competitor 1: Two knockdowns.
b Competitor 2: One knockdown and a first refusal.
c Competitor 3: Clear round.
d Competitor 4: Third refusal.
e Competitor 5: One knockdown and a foot in the water.
f Competitor 6: Second refusal and three knockdowns.
g Who would have won this section?
h If competitors have the same faults result, how are the placings decided?

3 In the dressage competition, riders make their horses perform a variety of disciplines and are awarded points out of ten for each skill.

Judge 1	Competitor	Skill 1	2	3	4	5	Total
	1	8.5	8.1	9.0	8.6	8.7	
	2	7.9	7.8	7.9	8.0	7.8	
	3	9.1	9.0	9.2	9.2	9.3	
Judge 2	**Competitor**						
	1	8.4	8.2	8.9	8.5	8.6	
	2	7.8	7.8	7.9	7.9	7.8	
	3	9.1	9.1	9.2	9.3	9.4	
Judge 3	**Competitor**						
	1	8.5	8.0	8.9	8.6	8.6	
	2	7.8	7.9	7.8	7.9	7.7	
	3	9.2	9.1	9.2	9.3	9.5	

a What was the total awarded by each judge to each competitor?
b What was the total awarded to each competitor? (a + d + g and so on)
c Who would have won the dressage and by how much?
d How much short of a perfect score was the winner?

ISBN: 9780170217095

Rowing

Did you know ...

- New Zealand's first rowing medal (bronze) was won at the 1920 Antwerp Games by Darcy Hadfield in the single sculls.
- Rowing began as an Olympic sport in 1900. It became an Olympic competition for women in 1976.
- Coxless means there is no extra person steering the boat.
- Steve Redgrave of Great Britain is the first rower to win four consecutive gold medals in the coxless pairs.

ISBN: 9780170217095

Exercise 20

1 Use the table of New Zealand's Olympic rowing results to answer the questions that follow. (G = gold, S = silver, B = bronze)

	Men								**Women**					
	Single sculls	**Double sculls**	**Coxless pairs**	**Coxed pairs**	**Coxed quads sculls**	**Coxless fours**	**Coxed fours**	**Coxed eights**	**Single sculls**	**Double sculls**	**Coxless pairs**	**Coxless quads sculls**	**Coxless fours**	**Coxed eights**
1920	B													
1932		S												
1968						G								
1972						S		G						
1976								B						
1984						G	B							
1988	B					B					B			
1992														
1996														
2000	G													
2004										G				
2008	B		B							G				

a In what years did New Zealand have most success at rowing?
b Which event have New Zealanders been best at?
c Which Olympic Games was the most successful for New Zealand? Explain your answer.
d How many rowers are there in a ...
i coxed fours?
ii coxless pair?
iii coxed eights?
e How many gold, silver and bronze medals have New Zealand rowers won in the past three Olympics?
f If New Zealand took a team for each event, how many rowers would be ...
i men?
ii women?
iii How many in total?

2 In 1920 Darcy Hadfield won his bronze medal in a time of 7:48.0, while Rob Waddell won gold at the 2000 Sydney Games with a time of 6:48.90. How much faster was Rob Waddell?

3 Copy the following table of gold medallist rowing times from the 2008 Beijing Games and use it to answer the questions that follow.

ISBN: 9780170217095

	Men		Women	
	Time	Rounded to nearest second	Time	Rounded to nearest second
Single sculls	6.59.83		7.22.34	
Double sculls	6.27.77		7.07.32	
Coxless pairs	6.37.44		7.20.60	
Coxed pairs	5.47.76		-	
Coxless eights	6.06.57		-	
Coxless fours	5.23.89		6.05.34	
Coxed fours	5.41.33		6.16.06	
Coxed eights	6.10.99		6.54.74	

- **a** Round each time to the nearest second.
- **b** Draw a bar graph to illustrate these times and compare the difference between men and women.
- **c** Which event has the biggest range between men and women?
- **d** Write 2–4 sentences explaining what you interpret from the graph.

4 All rowing races are over a straight 2000 m course.

- **a** How many kilometres is this?
- **b** If a single sculler finished a race in seven minutes, what was the speed in m/s? (answer to one decimal place)
- **c** The coxed eights want to complete the 2000 m in 5 min 48 sec. What would the times need to be at these stages?
  - **i** 500 m
  - **ii** 1000 m

5 Use this bar graph of boat lengths to answer the questions that follow.

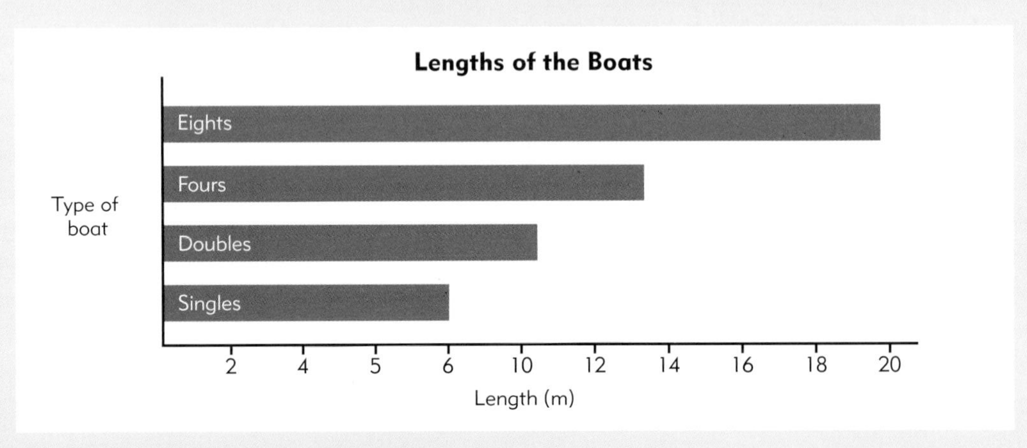

- **a** What is the length of each boat type?
- **b** Assuming each rower requires one metre of space, how much of the eights boat is occupied by rowers?
- **c** Using your answer to b, what percentage is this of the boat length? (answer to one decimal place)

ISBN: 9780170217095

6 Use this bar graph of boat weights to answer the questions that follow.

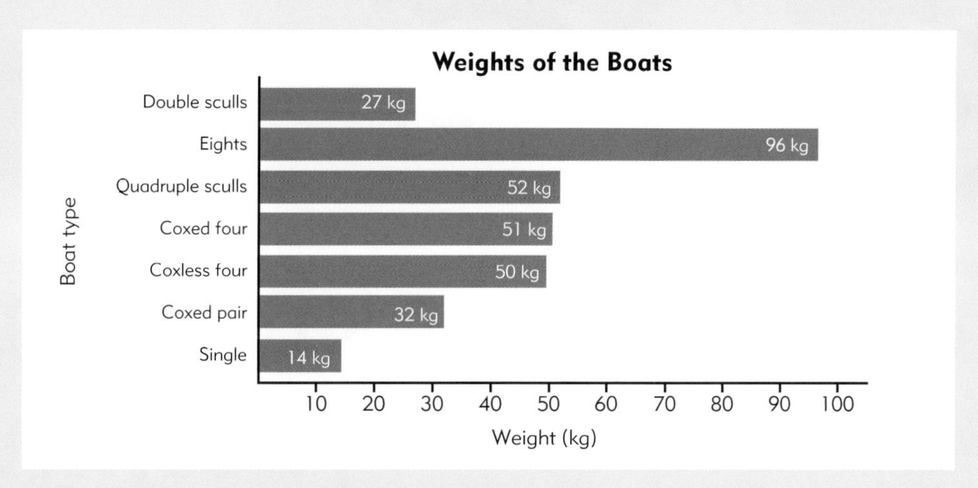

a The average weight of the coxless four rowers was 82 kg. What is the total weight of the rowers and boat?

b What is the total weight of rowers and boat for an eight crew whose weights are: 75 kg, 84 kg, 78 kg, 78 kg, 82 kg, 82.5 kg, 85 kg, 83.5 kg and 50 kg?

c What is your answer to b in tonnes?

d If the total weight of the coxed four and boat was 371 kg and the weight of the rowers was 68 kg, what was the weight of the cox?

e The double sculls want to have a total weight of no more than 200 kg. What is the maximum average weight each rower can be?

f The mean weight of the eight rowers is 82 kg, and when the weight of the cox is added to the total weight (crew plus cox) the mean weight becomes 79 kg. What is the weight of the cox?

ISBN: 9780170217095

Kayaking and yachting

Did you know ...

- There are two types of canoe: the kayak and Canadian canoe.
- Kayak competitors use a paddle with a blade at each end. Canadian canoeists use a paddle with a single blade.
- Canoeing was first introduced into the Olympics at Berlin in 1936.
- Yachting has been an Olympic sport since 1896. There are seven boat classes in yachting, which now includes boardsailing.

ISBN: 9780170217095

Exercise 21

1 Use the New Zealand kayaking Olympic medal table below to answer the questions that follow. (G = gold, S = silver, B = bronze)

	Kayak 1 500 m	Kayak 2 500 m	Kayak 2 1000 m	Kayak 4 1000 m
2008				
2004			S	
2000				
1996				
1992				
1988	B	G	S	
1984	G	G	G	G

a Explain what these events mean:
- **i** Kayak 1 500 m
- **ii** Kayak 2 1000 m
- **iii** Kayak 4 1000 m

b How many gold medals in total have New Zealand won in the Olympics?

c One kayaker has won four gold medals. What is his name?

d How many canoeists won medals in 1988?

2 Use these gold medallist times from the 2008 Beijing Games to answer the questions that follow.
- Kayak 1 500 m 1:37.252
- Kayak 2 500 m 1:28.736
- Kayak 2 1000 m 3:11.809
- Kayak 4 1000 m 2:55.714

a How much faster were the two kayakers over 500 m as compared with one?

b **i** If the kayak 2 500 m pair had paddled 1000 m, what would their time have been?

ii Would they have beaten the kayak 2 1000 m pair?

iii What is the difference in the times?

ISBN: 9780170217095

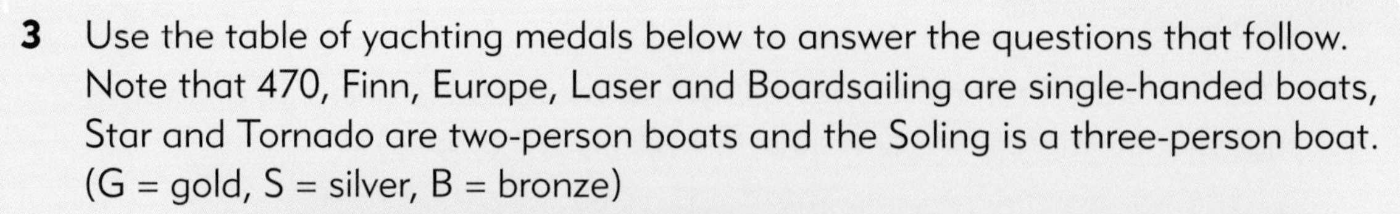

3 Use the table of yachting medals below to answer the questions that follow. Note that 470, Finn, Europe, Laser and Boardsailing are single-handed boats, Star and Tornado are two-person boats and the Soling is a three-person boat. (G = gold, S = silver, B = bronze)

	Men			Women			Mixed			
	470	Finn	Board sailing	470	Europe	Board sailing	Soling	Star	Laser	Tornado
1956										
1964										
1984			B							G
1988		B	G							S
1992		B		S		G		S		
1996						S				
2000			B			B				
2004										
2008			G							

a How many gold medals in total has New Zealand won in the Olympic Games since 1956?

b What year was our most successful? Explain your answer.

c If New Zealand took entrants for each class, how many sailors would be in the team?

4 Refer to the diagram of a yachting course below and answer the questions that follow.

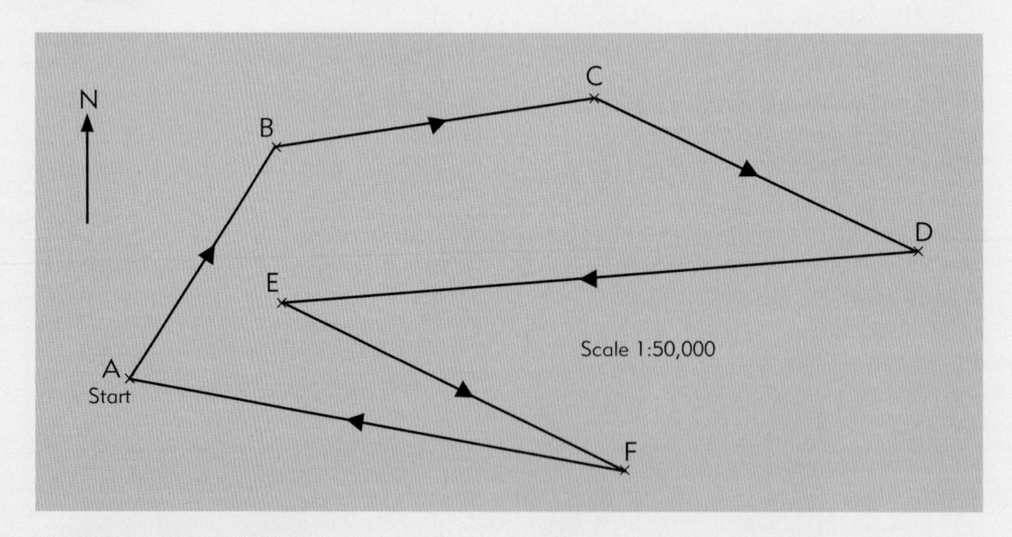

a What is the scale?

b What is the distance around the course?

c What are the compass directions at points A, B, C, D, E and F?

d Which point is approximately in a westerly direction?

e Which is the shortest leg of the course?

f Between which two points is halfway?

ISBN: 9780170217095

17 Other sports

New Zealand will also have competitors in archery, badminton, beach volleyball, boxing, cycling, hockey, judo, shooting, tennis and the triathlon.

Exercise 22

Choose one of the sports mentioned or illustrated above.

1 Find out as much as you can about the history of the sport.

2 Write two questions related to the sport involving some mathematical skills.

3 Prepare a presentation of your research.

ISBN: 9780170217095

18 New Zealand's past achievements

Did you know ...

- D'Arcy Hadfield (1920) won New Zealand's first medal.
- Ted Morgan (1928) won New Zealand's first gold medal.
- Yvette Williams (1952) won New Zealand's first female gold medal.
- Peter Snell (1960, 1964) was New Zealand's first triple gold medallist
- Ian Ferguson (1984) was New Zealand's first triple gold medallist at a single Games.
- Barbara Kendall (1992, 1996, 2000) became New Zealand's first female triple medallist.
- Valerie Adams is the current World, Olympic and Commonwealth champion for the women's shot put. Her personal best distance is 21.07 m. She won gold at Beijing.
- Mark Todd (1984, 1988) won Equestrian gold medals and has compted along with Andrew Nicholson in six Olympic Games. He could represent New Zealand at the 2012 London Olympics.

ISBN: 9780170217095

New Zealand's medal achievements in past Olympics

Venue	NZ athletes	Gold	Silver	Bronze	Total
1908 London	3	0	0	1	1
1912 Stockholm	3	1	0	1	2
1920 Antwerp	4	0	0	1	1
1924 Paris	4	0	0	1	1
1928 Amsterdam	10	1	0	0	1
1932 Los Angeles	21	0	1	0	1
1936 Berlin	7	1	0	0	1
1948 London	7	0	0	0	0
1952 Helsinki	15	1	0	2	3
1956 Melbourne	52	2	0	0	2
1960 Rome	38	2	0	1	3
1964 Tokyo	69	3	0	2	5
1968 Mexico City	59	1	0	2	3
1972 Munich	96	1	1	1	3
1976 Montreal	86	2	1	1	4
1980 Moscow	4	0	0	0	0
1984 Los Angeles	134	8	1	2	11
1988 Seoul	93	3	2	8	13
1992 Barcelona	144	1	4	5	10
1996 Atlanta	97	3	2	1	6
2000 Sydney	151	1	0	3	4
2004 Athens	151	3	2	0	5
2008 Beijing	166	3	2	4	9
Total	1414	37	16	36	89

Exercise 23

Use the above table to answer these questions.

1 What has been our largest team, and to which Olympic Games were they sent?

2 What is the total number of athletes sent to all Games?

3 What is the total number of medals won at all Games?

4 Using your answer in question 3, what is the percentage of medals won relative to the number of athletes participating? (answer to one decimal place)

5 Using three points for a gold medal, two for silver and one for bronze, which Games were New Zealand's ...

a best? **b** worst?

6 Compare the number of athletes with the total number of medals won.

a Which was New Zealand's best Games?

b Using your answer to 6a, what percentage of athletes won medals at these Games?

c Use this method to calculate New Zealand's best six years.

7 **a** Draw a pie graph to illustrate the total gold, silver and bronze medals won in all Olympic Games. (Hint: $360° \div 89 = 4.04$, so use 4.)

b Write 2–4 sentences about the graph.

ISBN: 9780170217095

8 a How many Olympic Games has New Zealand sent athletes to?
b What is the average number of athletes sent to the Games?
c What is the average number of medals won?

9 Use the table below of New Zealand's medal achievements in individual sports to answer the questions that follow.

New Zealand's medal achievements in individual sports

	Medals			
	Gold	Silver	Bronze	Total
Archery				
Athletics	9	2	9	20
Badminton				
Beach volleyball				
Boxing	1	1	1	3
Canoeing	5	2	1	8
Cycling	1	1	2	4
Equestrian	3	2	4	9
Fencing				
Hockey	1			1
Judo				
Rowing	6	2	8	16
Shooting			1	1
Swimming	3	1	3	7
Table tennis				
Tennis			1	1
Triathlon	1	1	1	3
Wrestling				
Yachting	7	4	5	16
Total	37	16	36	89

a What is our most successful sport?
b i Rank the top six sports for New Zealand at the Olympic Games.
ii Write 2–4 sentences about these sports and why you think New Zealanders are more successful at them.
c i Draw a bar graph to illustrate the medals won. Choose a different colour for gold, silver and bronze and only use those sports that have been successful.
ii Write 2–4 sentences about your graph.

10 Select a topic that investigates New Zealand's past achievements at the Olympic Games. This could be a specific sport, a particular venue or the achievements of a certain athlete. Plan and conduct the investigation using the statistical enquiry cycle.

ISBN: 9780170217095

Olympic trivia

Did you know ...

- The oldest male Olympic competitor was Oscar Swahn (SWE), aged 72 years 279 days, for shooting in 1920. The oldest female was Lorna Johnstone (GBR), aged 70 years five days, for equestrian in 1972.
- The youngest male Olympic competitor was an unnamed French youth, aged between 7–10 years, for rowing in 1900. The youngest female was Liana Viceris (PUR), aged 11 years 328 days, for swimming in 1968.
- The most Olympic medals won by an individual male and female competitor are Nikolay Andrianov (URS), who won 15 in 1972–82 for gymnastics, and Larissa Latynina (URS), who won 18 medals for gymnastics 1956–64.

ISBN: 9780170217095

Countries and athletes attending the Games

Olympic year		Host city	Nations	Women	Men	Total
I	1896	Athens	13	-	311	311
II	1900	Paris	22	12	1318	1330
III	1904	St Louis	13	8	617	625
IV	1908	London	22	36	2020	2056
V	1912	Stockholm	28	55	2491	2546
VII	1920	Antwerp	29	64	2628	2692
VIII	1924	Paris	44	136	2956	3092
IX	1928	Amsterdam	46	290	2724	3014
X	1932	Los Angeles	37	127	1281	1408
XI	1936	Berlin	49	328	3738	4066
XIV	1948	London	59	385	3714	4099
XV	1952	Helsinki	69	518	4407	4925
XVI	1956	Melbourne	67	371	2813	3184
XVII	1960	Rome	83	610	4736	5346
XVIII	1964	Tokyo	93	683	4457	5140
XIX	1968	Mexico City	112	781	4749	5530
XX	1972	Munich	122	1010	6086	7156
XXI	1976	Montreal	92	1251	4834	6085
XXII	1980	Moscow	81	1088	4238	5326
XXIII	1984	Los Angeles	140	1620	5458	7078
XXIV	1988	Seoul	159	2186	6279	8465
XXV	1992	Barcelona	169	2707	6657	9364
XXVI	1996	Atlanta	196	4000	6788	10 788
XXVII	2000	Sydney	199	4069	6582	10 651
XXVIII	2004	Athens	201	4329	6296	10 625
XXVIX	2008	Beijing	204	4637	6305	10 942
XXX	2012	London	c.205	?	?	c.17 000

Exercise 24

1 Why were there no Games in 1916, 1940 or 1944?

2 **a** What percentage of the athletes at the 1900 Paris Games were women? (round to nearest whole number)

b Compare this percentage with the percentage of women attending the 2008 Beijing Games. Explain your answer.

3 What percentage of men attended the 1952 Helsinki Games and 2008 Beijing Games? (answer to one decimal place)

4 If a similar percentage of women athletes attend the 2012 London Games, how many would you expect to attend?

ISBN: 9780170217095

5 **a** What is the percentage increase in nations attending the Games from 1896 to 2008?
b What is the percentage increase in total number of athletes attending the Games from 1896 to 2008? (answer to one decimal place)

6 At the 2008 Beijing Olympic Games ...
a What was the average number of athletes attending per country? (round sensibly)
b What was the average number of:
i men attending per country?
ii women attending per country?
(How can you check your answer?)

7 Use the medal table below for five countries from three Olympic Games (2000–2008) to answer the questions that follow.

	Medals				Population
	Gold	Silver	Bronze	Total	
New Zealand	10	5	9	24	4.2 million
Australia	56	65	73	195	17 million
Great Britain	40	40	40	120	55 million
United States of America	152	133	254	539	241 million
People's Republic of China	127	76	69	272	1.3 billion

a Calculate the ratio of total medals won to the population of each country.
b Rank the countries in order, using your answer in **a**.
c Which country has the best ratio?
d If New Zealand had the same population as Australia and continued to win medals at this rate, how many medals would we have won?
e **i** Calculate the ratio of gold medals won to the total number for each country.
ii Which country has the best ratio?

8 Funding and organising the Games is a major exercise. The 1896 Athens Games were funded by stamps, ticket sales, commemorative medals, programme advertising and private donations. Today it is funded by the sale of television rights, sponsorship licensing, ticket sales, government and city contributions, coins and stamps. In 2008 the Games cost more than $43 billion. London 2012 is expected to cost £5 billion.

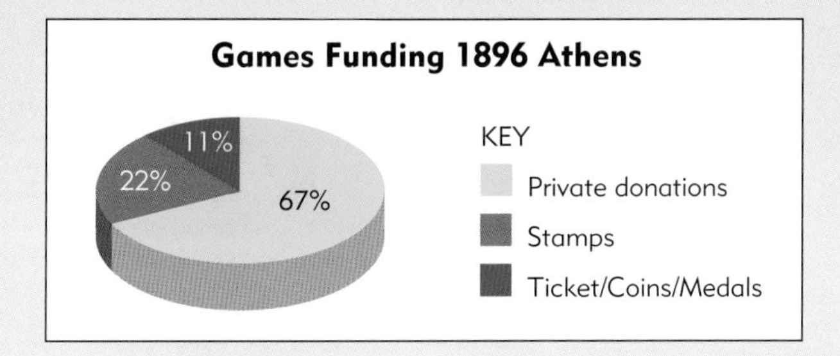

ISBN: 9780170217095

Games funding plan for London 2012
Lotteries: £1.5 billion
Council tax: £625 million
London development agency: £250 million
IOC revenues: £560 million
Official suppliers and sponsorship: £450 million
Broadcasting rights: £2.5 billion
Ticket sales: £300 million
Licensing revenues: £60 million

- **a** Construct a pie chart to show how London plans to fund their Games in 2012.
- **b** Which form of funding provided the most income in 1896?
- **c** If the total funding required for the Athens Games in 1896 was USD$2000, how much came from ...
  - **i** private donations?
  - **ii** stamps?
  - **iii** tickets, coins, medals?
- **d** What do you think were included in the other category?

9 The table below compares the 1896 Olympic Games in Athens with the 2012 London Games.

	Athens 1896	London 2012
Days	5	19
Sports	9	26
Events	32	302
Countries	13	205
Athletes	311	17 000
Tickets available	c. 60 000	9 000 000

- **a** What is the percentage increase (answer to one decimal place) from 1896 to 2012 for:
  - **i** number of days?
  - **ii** number of events?
  - **iii** number of countries?
  - **iv** number of tickets available?
- **b** Which has the greatest increase? Explain your answer in 2–3 sentences.
- **c** Compare the number of athletes participating with the number of days. Which year has the greater participation?

10 Each country produces postage stamps at the time of the Games. These are New Zealand's stamps from the 2008 Beijing Olympics.

ISBN: 9780170217095

a Research if there is a maximum or minimum size for a stamp, and what shapes they can be.

b Design and construct a series of stamps suitable for the London 2012 Games.

c What value of stamps will be needed?

d Copy this table. Using the selection of 2008 Beijing Olympic stamps supplied, complete the column 'Stamps required'. (Remember to use as few stamps as possible.)

Value of article to be posted	Stamps required
$1.20	
$1.90	
$4.50	
$7.20	
$10.00	

11 The table below outlines the postage rates for New Zealanders. (All prices include GST.) Use this table to answer the questions that follow.

Size	Maximum dimensions per letter (height x length)	Maximum weight per letter	Maximum thickness per letter	Required postage per letter for Australia and South Pacific	Required postage per letter for Rest of the World
Aerogrammes and postcards	130 mm x 235 mm	10 g		$1.90	$1.90
Medium	130 mm x 235 mm	200 g	10 mm	$1.90	$2.40
Large	165 mm x 235 mm	200 g	10 mm	$2.40	$2.90
Extra large	230 mm x 325 mm	200 g	10 mm	$2.90	$3.40
Oversize	260 mm x 385 mm	200 g	10 mm	$3.60	$5.10

a How much will it cost to send three postcards and two 230 mm x 325 mm letters to London?

b Jane needs to send a 260 mm x 385 mm letter weighing 150 g to London. What will it cost and how much change will she get from $10?

c Calculate the increase in area between a postcard and an oversized letter. How much extra does this oversized letter cost to post to London?

ISBN: 9780170217095

12 Below is a small sample of postage rates in the UK. Use this table to answer the questions that follow.

Airmail letters to Europe		Airmail letters to rest of the world
Weight	**Postage**	**Postage**
Postcard	68 p	76 p
10 g	68 p	76 p
60 g	£1.30	£2.07
100 g	£1.73	£3.19
200 g	£3.11	£5.92

a What would it cost in British £ to send five postcards to New Zealand and three postcards to Australia?

b I want to send a 150 g letter from London to New Zealand.

i How much will it cost in British £?

ii How much will that be in NZ$, if GDP£1 = NZ$2.09222?

c Compare the cost of sending a postcard from London to New Zealand with sending a postcard from New Zealand to London. Use the exchange rate in b ii, above. Which was is cheaper?

d Compare the rate per gram for a 10 g letter with a 200 g letter. How does this compare with the rate per gram for a 10 g and medium 200 g letter posted in New Zealand. Is postage cheaper in New Zealand or Great Britain?

13 Select one of the Olympic facts from the box on the following page and calculate and/or research something about it. For example:

How long would it take to make the 17 320 beds in the Olympic Village?

or

How many cases of apples are required for the duration of the Games at the Olympic Village?

ISBN: 9780170217095

Olympic facts from the 2008 Beijing Games

Did you know ...

- The 2008 Beijing Olympics opened at 8 minutes past 8 on 8 August 2008, because the number 8 is a very special number and signifies prosperity in Chinese culture.
- 10 308 athletes, 20 000 officials and 21 600 media workers participated.
- BOCOG recruited 70 000 volunteers.
- The torch relay that started in Athens on 24 March 2008 took 130 days, covered 137 000 km and involved 21 880 torch bearers. The torch was designed to remain lit in a 65 km/h wind, at -40°C or in rain up to 50 mm per hour.
- It cost China US$43 billion to stage the 2008 Games.
- The Games attracted a worldwide television audience of 4.7 billion people. One billion watched the opening ceremony on television.
- Beijing Olympics Broadcasting – the official broadcasting company – produced 5400 hours of programming, 2000 hours more than Athens.
- There were 302 events in 28 sports at 31 venues in Beijing, and at six venues outside Beijing. There were 302 gold medals decided at Beijing (compared to 301 in the Athens Games).
- Seven million tickets were available for the Games.
- The National Stadium (Bird's Nest) has a seating capacity of around 98 000, and the National Aquatic Centre (The Cube) can seat 17 000.
- The opening ceremony cost US$1m to produce and featured more than 15 000 performers.
- At the closing ceremony, 2583 lights weighing over 300 tonnes were applied.
- Forty-three new World records and 132 new Olympic records were broken.
- Eighty-six countries won at least one medal.
- Michael Phelps of the USA won the most gold medals at a single Olympic Games and the most career gold medals, by winning eight gold medals.
- There were nine new events introduced. Some of these were BMX, 300 m steeplechase for women, and a 10 km open swim for men and women.
- The Olympic Village occupied 66 ha, with accommodation for 16 000 athletes and officials. The restaurants were open 24 hours a day for the duration of the Games, and could seat 5000 people at a time.
- Normal beds were 1 m x 2 m but many rooms were fitted with beds measuring 2 m x 4 m to accommodate the taller/bigger athletes.
- Four hundred million young Chinese pupils, from 500 000 schools around China, had lessons about the Olympics prior to the Games.
- Baseball and softball were included in the Olympics for the first time.
- The USA has 4.6% of the world's population, but won 11.5% of the medals. China has 19.8% of the world's population and won 10.4% of the medals, whereas India has 17% of the world's population yet only won 0.31% of the medals.
- 15 000 Chinese couples were married in Beijing on 8 August 2008.

ISBN: 9780170217095

The Paralympics

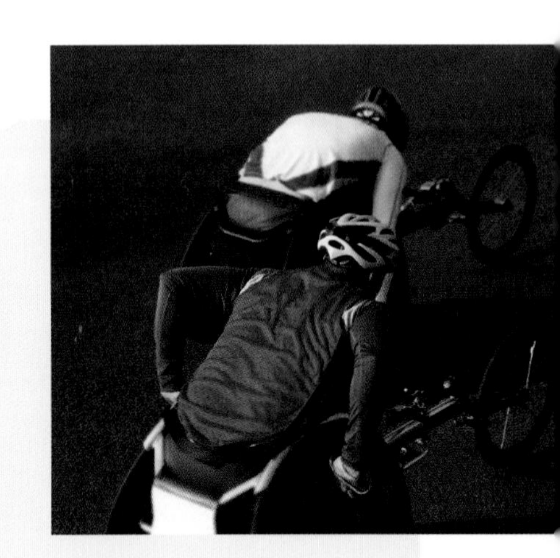

Did you know ...

- The Paralympics were first introduced as part of the Olympics in 1948.
- A small group of British veterans from the Second World War were the first group of participants.
- The Paralympic Games is now one of the largest international sports events.
- It is an international multi-sport event for athletes with physical disabilities.

Exercise 25

1 List some of the disabilities that athletes can have.

2 There are broad categories for each sport in which the athletes compete. What are the categories for swimming?

3 Select another sport and investigate the categories.

4 When are the Paralympics held and how often?

5 Are there summer and winter Paralympics?

6 In what year and in which country was the first Paralympics officially held?

7 A famous New Zealand Paralympian competed in both the Olympics and Paralympics in archery. What is the athlete's name and in what year did they compete in both Games?

8 As at 2008, an American athlete holds the title for the most successful Paralympian, winning 55 medals, 41 of which were gold. Who is this athlete, how long was their career and in what sport did they compete?

ISBN: 9780170217095

2008 Beijing Paralympics

Did you know ...

- The Beijing Paralympic Games were held from 6–17 September 2008.
- There were 4200 athletes from 148 countries competing in 471 events across 20 sports.
- The Paralympic Games used 20 of the Beijing Olympic Games venues.
- The Olympic Village closed on 27 August 2008 and reopened on the 30 August, fully refurbished for the Paralympians.
- There were 30 000 volunteers at these Games.

Exercise 26

1 What was the motto for the 2008 Beijing Paralympic Games?

2 Research the time and route taken for the Paralympics torch relay.

3 Sketch the design of the Paralympic flag.

4 List the 20 sports in the Beijing Paralympics.

5 List those sports that the New Zealand Paralympians participated in.

6 In 2008 a South African paralympian attempted to qualify for the 2008 Olympic Games. Who is this athlete, what is their disability and in which events did they compete?

7 Another South African athlete became the first athlete to carry their country's flag for both the Olympic and Paralympic opening ceremonies. What was this athlete's name and in which sport did they compete?

8 Who was the youngest swimmer ever to win a swimming gold medal in Beijing?

9 Boccia is a sport at the Paralympics.

a Research two characteristics about this sport.

b Name the five countries that shared the seven gold medals.

10 Below is a table of some of the sports at the 2008 Beijing Paralympic Games with the number of competition days, the number of athletes and the number of gold medals available for that sport.

Sport	**Number of competition days**	**Number of athletes**	**Number of gold medals available**
Athletics	10	1035	160
Swimming	9	560	141
Equestrian	6	70	11
Wheelchair rugby	5	96	1
Wheelchair tennis	8	112	6
Wheelchair basketball	10	264	1
Rowing	3	96	4
Archery	7	136	9
Shooting	6	140	12
Cycling	7	188	44

ISBN: 9780170217095

a On average, how many athletes in these sports competed in each day of competition? (answer to nearest whole number)
- **i** Athletics
- **ii** Swimming
- **iii** Rowing
- **iv** Wheelchair basketball

b What is the ratio of gold medals available to the number of athletes in these sports?
- **i** Equestrian
- **ii** Wheelchair tennis
- **iii** Archery
- **iv** Cycling

c Of the four sports in **10b**, which sport gives you a better chance of winning a gold medal?

d The Paralympics competition started on 7 September and finished on 17 September 2008. What percentage of competition days were these sports held? (answer to one decimal place)
- **i** Shooting
- **ii** Athletics
- **iii** Wheelchair rugby
- **iv** Rowing

e There were 4200 athletes competing at the Games. What fraction of athletes competed in these sports? (answer to the nearest whole number)
- **i** Athletics
- **ii** Swimming
- **iii** Wheelchair basketball
- **iv** Rowing

11 Use the 2008 Beijing Paralympics medal table below to answer the questions that follow.

2008 Beijing Paralympics Medal Table

Rank	Nation	Number of athletes	Gold	Silver	Bronze	Total
1	China	332	89	70	52	211
2	Great Britain	212	42	29	31	102
3	United States	213	36	35	28	99
4	Ukraine	243	24	18	32	74
5	Australia	170	23	29	27	79
6	South Africa	63	21	3	6	30
7	Canada	143	19	10	21	50
13	South Korea	79	10	8	13	31
24	New Zealand	30	5	3	4	12
58	Argentina	42	0	1	5	6
69	Syria	8	0	0	1	1
	Total (for 69 countries)		**473**	**471**	**487**	**1431**

ISBN: 9780170217095

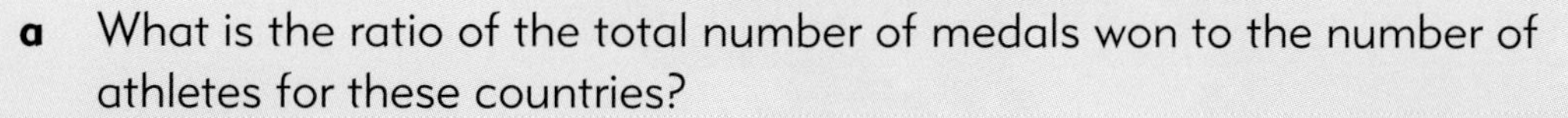

a What is the ratio of the total number of medals won to the number of athletes for these countries?
- **i** China
- **ii** Australia
- **iii** South Africa
- **iv** New Zealand

b What is the ratio of gold medals won to the number of athletes for these countries?
- **i** Great Britain
- **ii** Canada
- **iii** New Zealand
- **iv** Ukraine

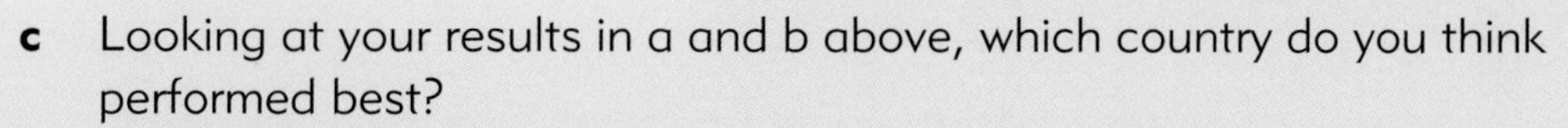

c Looking at your results in a and b above, which country do you think performed best?

d What percentage of the gold medals available did these countries win?
- **i** China
- **ii** Great Britain

Exercise 27

1. What are the dates of the 2012 London Paralympic Games?
2. List the sports being offered at the 2012 London Paralympic Games.
3. How many tickets will be available?
4. How many nations are expected to participate?
5. Which events and how many athletes are New Zealand planning to send to the 2012 London Paralympic Games?
6. Research one of the Paralympic sports being offered at London. Find out what the categories of disabilities are, and what differences there are between this sport and its able-bodied equivalent.
7. The 2012 London Paralympic Games is introducing another disability for the first time. What is it?
8. Select a topic that investigates New Zealand's past achievements at the Paralympic Games. This could be a specific sport, a particular Games or the achievements of a certain athlete. Plan and conduct the investigation using the statistical enquiry cycle.

ISBN: 9780170217095

Putting it into practice

Achievement Standard 91030: Apply measurement in solving problems

Internal Assessment • 3 credits

GM 6:2 Apply the relationships between units in the metric system, including the units for measuring different attributes and derived measures.

GM 6:3 Calculate volumes, including prisms, pyramids, cones and spheres, using formulae.

1 **a** Convert 10 m to cm.
b How many km in 5695 cm?
c Convert 6.5 km to cm.

2 **a** Convert 5 kg to g.
b How many kg in 1234 g?
c Convert 453 mg to g.

3 **a** How many minutes in 3 and ¾ hours?
b How many seconds in 2½ minutes?
c How many hours and minutes in 167 minutes?

4 **a** If a sprinter can run 100 m in 10 sec, what is the speed in m/s?
b If a marathon runner can run 10 km in 35 min, what is the time for a marathon of 42 km?
c If a female runner can run the outdoor 5000 m in 14:11:15 and a male runner 12:37:35, who is the faster runner and by how much?

5 The men's Olympic shot put weighs 7.260 kg and the womens weighs 4 kg.
a What is the weight difference in grams between men and women?
b The men's discus weighs 2 kg and has a diameter of 219 – 221mm. What is this weight in pounds?

ISBN: 9780170217095

c The women's discus is 2 lb 3 oz in weight and has a diameter of 7.17 in. What is the diameter in mm?

6 The '6 metres club' in the men's pole vault is an elite group of athletes who have cleared 6 m. Sergey Bubka of the Ukraine heads the outdoor list with a vault of 6.15 m in 1985.

a What is the difference between the outdoor and indoor heights in cm and inches?

b In 2009 Renaud Lavillenie of France vaulted 6.01 m outdoor and 6.03 m indoor. What is the difference in inches?

c The 50 m freestyle swim can be swum over a short course (2x25 m) or long course (1x50 m). The men's time for the short course is 20.30 sec and the long course 20.91 sec. The women's times for the short course is 23.73 sec and for the long course 23.25 sec. What is the speed in m/s for the men and women over the long course?

7 If the length of an eights rowing boat is 20 m, a fours boat is 13.5 m and a doubles boat is 10.5 m long.

If each rower requires 1 m of space, how much of these boats is not occupied by the rowers?

- **i** coxed eights
- **ii** coxless fours
- **iii** double sculls

b Rowing has special weight categories. Lightweight for men is an average weight of 70 kg with no rower over 72.5 kg, and women's average weight of 57 kg with no rower over 59 kg.

- **i** In the men's coxed eights, a lightweight crew has an average of 71 kg for 8 rowers. What is the maximum weight of the cox?
- **ii** In the women's coxless fours, a lightweight crew has three weights of 48 kg, 60 kg, 55 kg. What is the maximum weight of the 4th rower?

c The average weight of a women's coxed eight is 63kg. The weights of 7 of the rowers was 68 kg, 64 kg, 59 kg, 68 kg, 63 kg, 69 kg, 62 kg and the cox is ⅔ of the weight of the 8th rower. What is the weight (to the nearest whole number) of the cox?

8 a The swimming pool is 25 m x 10 m x 3 m. What is its volume?

b How many litres of water does it hold?

c If the pool is 25 m long, 10 m wide, 4 m deep at the deep end and 1.5m deep at the shallow end. What is the volume of water in this pool?

9 a The men's shot put weighs 7.26 kg. What is this weight in pounds?

b The shot put circle has a diameter of 2.135 m. What is the area of the circle?

c The shot put must land in the legal sector of 34.92° of the throwing area. What percentage is this of the whole circle?

10 The 110 m hurdles is for men and the 100 m is for women. For men ten hurdles of 1.067 m in height are evenly spaced over 110 m, while for women the hurdles are 0.838 m in height and are evenly spaced over 100 m.

a The first hurdle for the 110 m hurdles is placed at 13.72 m from the start. If the last hurdle is 14.02 m from the finish, what is the spacing between the other nine hurdles? (answer to 2 dp)

ISBN: 9780170217095

   b The first hurdle for the 100 m hurdles is placed 13 m from the start. If the last hurdle is 10.5 m from the finish, what is the spacing between the other nine hurdles? Answer to 2 dp)
   c The men's 400 m hurdles is run in 46.78 s by Kevin Young. What is this time in km/h?

11 The ratio of the hop, step and jump is 1:3:4.
   a What is the length of the hop of the men's world record holder, whose jump length is 18.29 m?
   b What is the length of the jump of the women's record holder, whose jump length is 15.50 m?
   c If an athlete has a step length of 1.98 m what would be her hop, step and jump length?

12 **a** If a stadium holds 100 000 people and 90 000 tickets are sold, what percentage are unsold?
   b If it is estimated that 14 500 people will attend the gymnastic final and actually only 13 900 people attend, what is the percentage error of the attendance? (answer to one decimal place)
   c The average ticket price is £85 for the Opening Ceremony and there are 80 000 tickets available. If the present exchange rate for NZ$ is £0.4590, how much income in NZ$ will come from the Opening Ceremony ticket sales?

13 The Smith family of 4 (2 adults, one 14-year-old, one 9-year-old) from New Zealand decide to go to London for the 2012 Olympic Games for 24 days. Using the following information and other information you can research, calculate the cost of their 24 day holiday in London. Set out your calculations clearly and logically, justifying your answers with the information used.

- Adult airfares one way. 12 – 16 year olds 75% of adult fare, 7 – 11 year olds 50% of adult fare.
- Accommodation - £152 per night for a two bedroom suite – no breakfast included.
- Tickets for Opening and Closing Ceremony £105 per person, no reduction for children.
- Tickets for 10 days of the competition at an average price (for adults and children) of £25 a ticket.
- Transport around London.
- Parking.
- Average meal costs per day.
- Average sightseeing costs per day.
- Incidentals.
- Shopping.
- Exchange rate £0.4585 = $1.
- Final costs need to be in New Zealand dollars.

ISBN: 9780170217095

Achievement Standard 91034: Apply transformation geometry in solving problems

Internal Assessment • 2 credits

GM 6:8 Compare and apply single and multiple transformations.
GM 6:9 Analyse symmetrical patterns by the transformations used to create them.

1 a Reflect this Polish flag in the y axis.

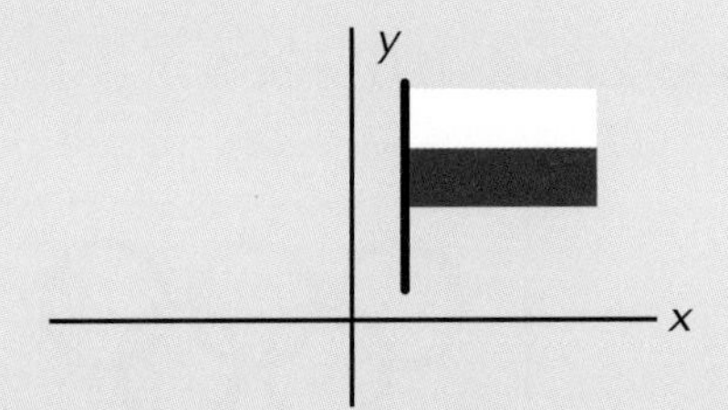

b Reflect this Japanese flag in the x axis.

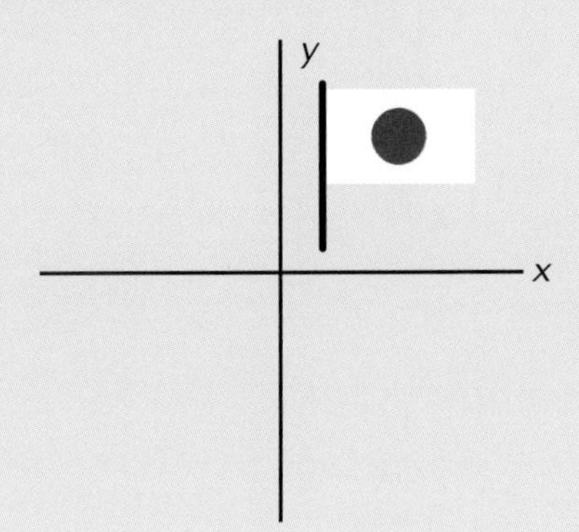

c Reflect this Olympic symbol in the y axis.

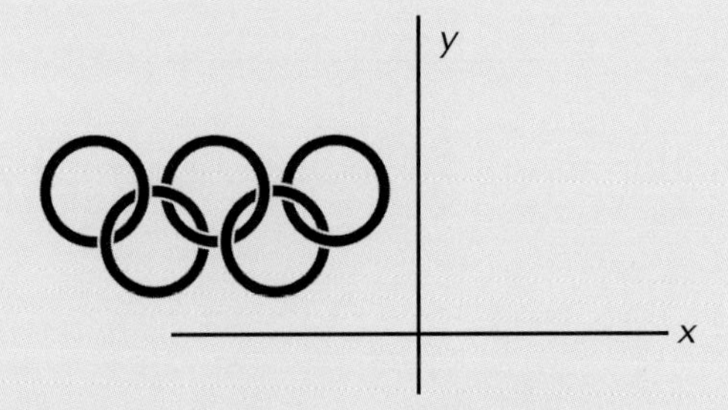

2 a Rotate this Greek flag about (0,0) 90° clockwise.

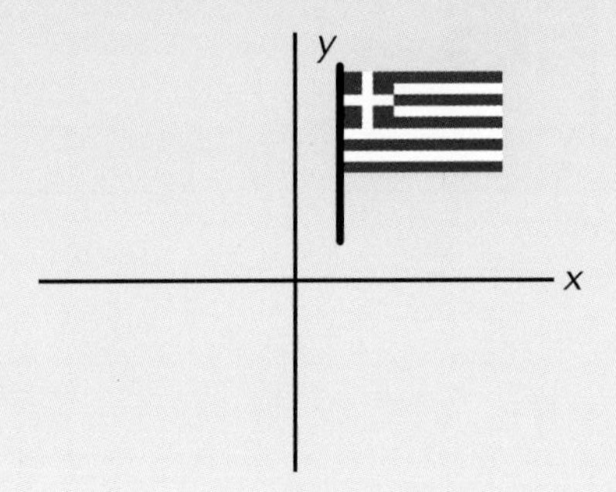

ISBN: 9780170217095

b Rotate this Peruvian flag about (0, 0) 270° anticlockwise.

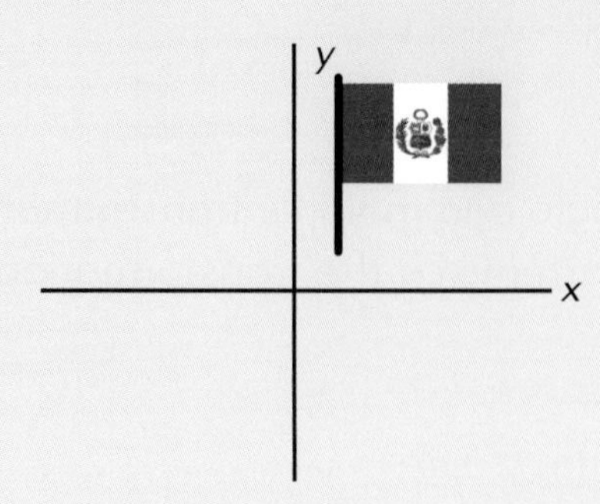

c Rotate this Olympic symbol about (0,0) 180°.

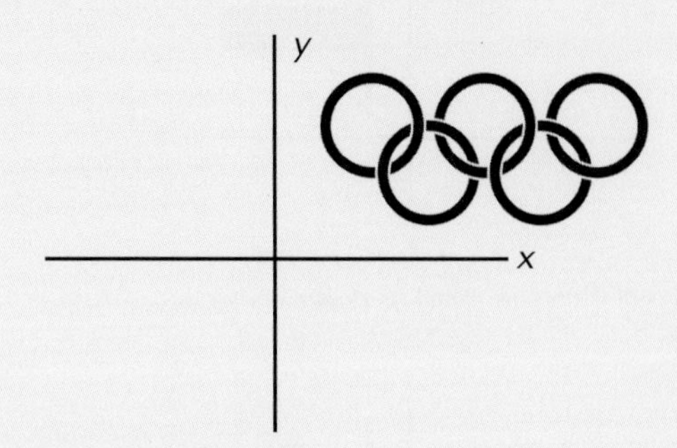

3 **a** Translate this Australian flag $\begin{pmatrix} 4 \\ 4 \end{pmatrix}$

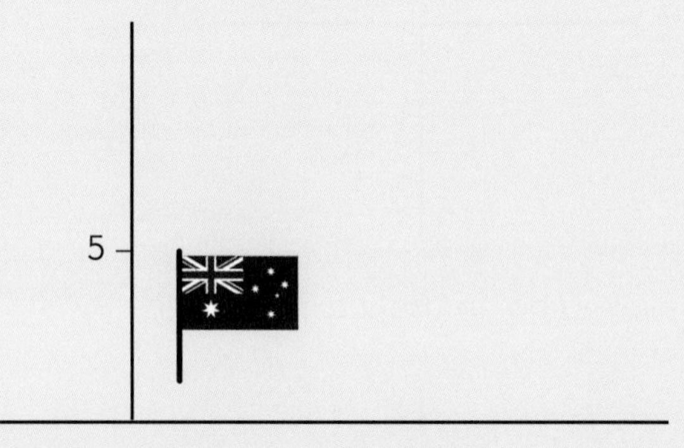

b Translate this United Kingdom flag $\begin{pmatrix} -6 \\ 0 \end{pmatrix}$ and colour correctly.

ISBN: 9780170217095

c Translate this Olympic symbol (posting a letter) $\begin{pmatrix}-5\\-6\end{pmatrix}$

4 a Enlarge this Tanzanian flag by a scale factor of 2 using (1, 1) as a centre of enlargement.

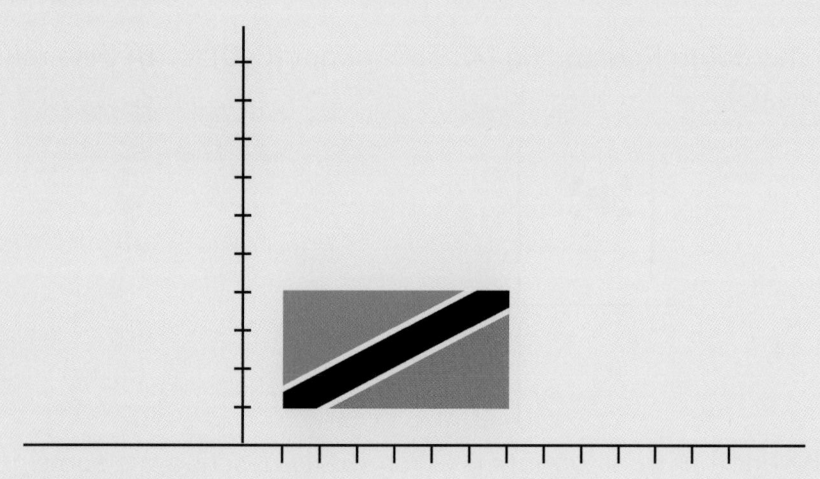

b Enlarge this New Zealand flag by a scale factor of -1 using (0, 0) as a centre of enlargement.

c Enlarge this Chinese flag by a scale factor of -2 using (0, 0) as a centre of enlargement.

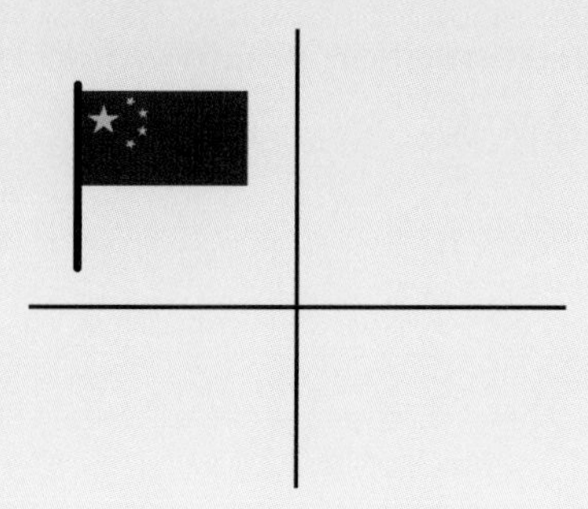

ISBN: 9780170217095

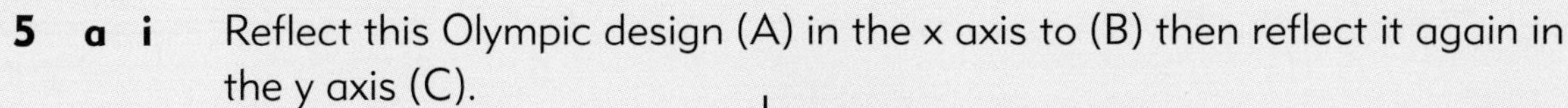

5 **a** **i** Reflect this Olympic design (A) in the x axis to (B) then reflect it again in the y axis (C).

ii What is another transformation which would describe the same transformation from A to C?

b **i** Rotate this South Korean flag (A) 180° about (0, 0) to (B) then reflect it in the y axis (C).

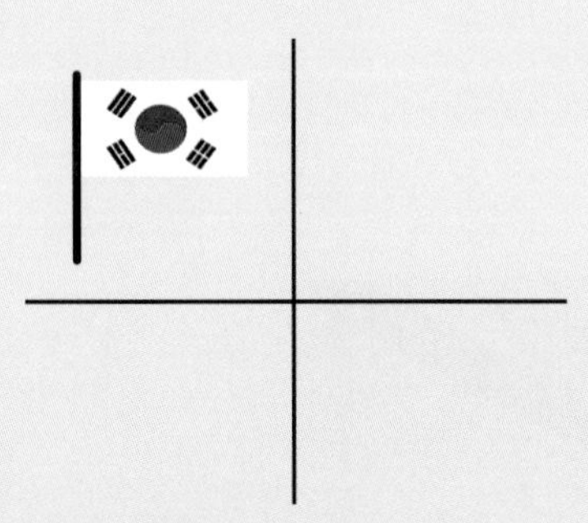

ii Describe another transformation which would describe the same transformation from A to C?

c **i** Enlarge this symbol (A) by a scale factor of 2, using (-1, 1) as a centre of enlargement (B) then rotate it about (0,0) 180° (C).

ii Describe another transformation which would describe the same transformation from A to C.

ISBN: 9780170217095

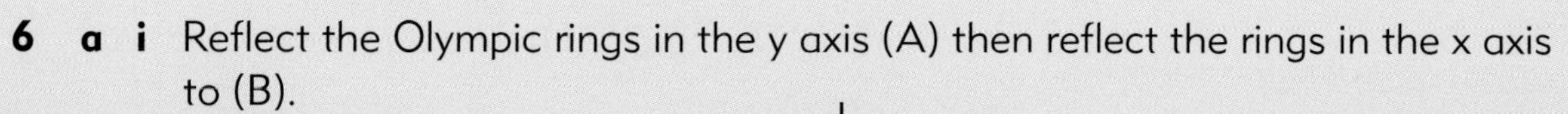

6 **a** **i** Reflect the Olympic rings in the y axis (A) then reflect the rings in the x axis to (B).

ii Describe fully the transformation require to get image B back to the original A.

7 **a** Describe fully the transformation the New Zealand flag has undergone from image A to B.

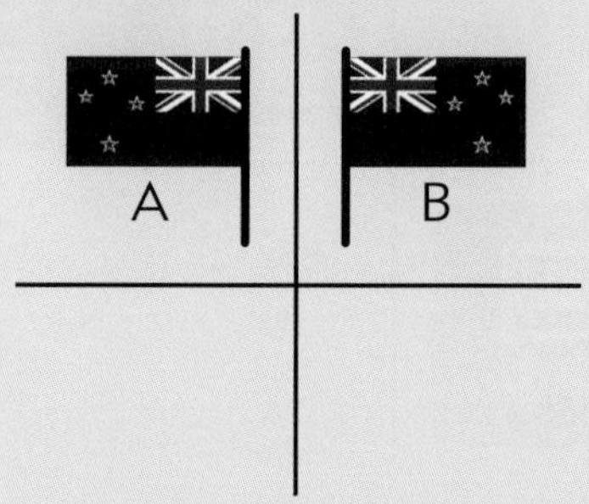

b Describe fully the transformation the New Zealand flag has undergone from image A to B.

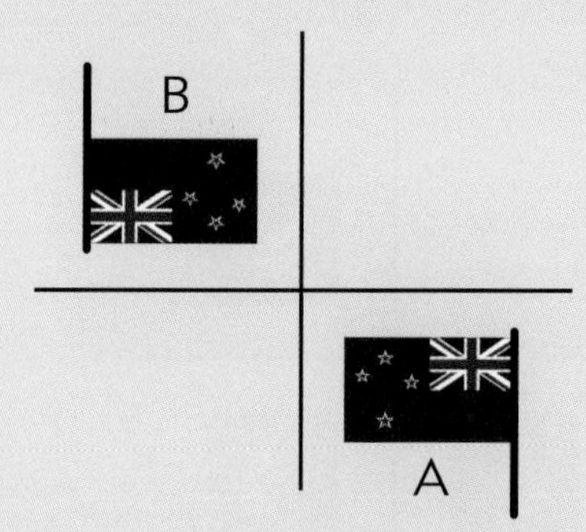

c Describe fully the transformation the New Zealand flag has undergone from image A to B to C.

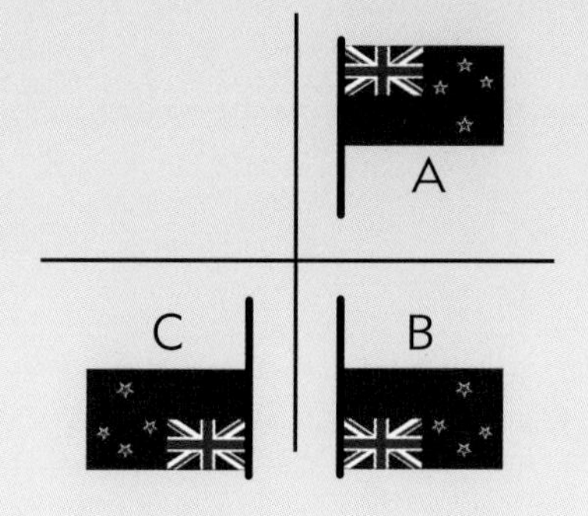

ISBN: 9780170217095

8 **a** Describe fully the transformation this Olympic symbol has undergone from A to B.

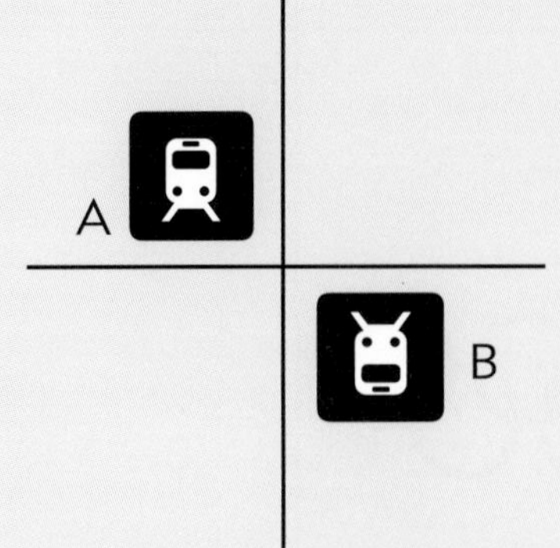

b Describe fully the transformation this Olympic Symbol has undergone from A to B to C.

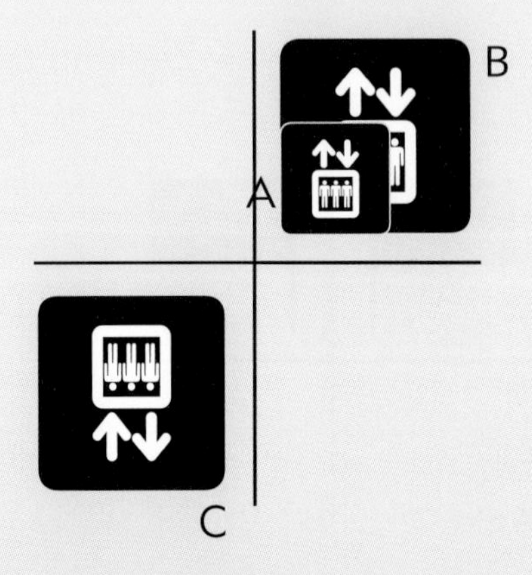

c Describe fully a two stage transformation which these Olympic symbols may have undergone to get from A to C.

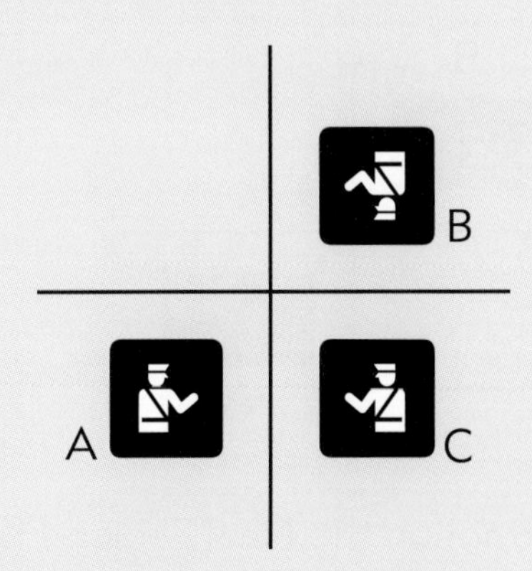

ISBN: 9780170217095

Enrichment

1. The athlete's uniforms – for competing and for official ceremonies – is an important aspect of the team. Design a uniform taking into consideration London's climate.
2. Every four years there is a winter Olympic Games. Research some aspect of these Games.
3. Sponsorship, advertising and television have a major impact on the financial aspect of hosting the Olympic Games. Investigate these issues and how they affect the New Zealand Olympic team, as well as either the Beijing Olympic Committee for the Olympic Games (BOCOG) or the London Olympic Committe for Olympics Games (LOCOG).
4. What effects have the 2008 Beijing Games had on the tourism industry in China?
5. Word search. Find these 20 sporting words.

archery	athletics
swimming	horse
judo	boxing
taekwondo	yachting
rowing	canoe
gymnastics	fencing
hockey	ball
cycling	tennis
weights	triathlon
oar	rocket

C	O	O	C	Y	C	L	I	N	G	R	G
A	R	C	H	E	R	Y	O	T	R	A	Y
N	T	A	E	K	W	O	N	D	O	O	M
O	U	A	V	C	Q	Z	U	H	W	P	N
E	P	S	T	O	R	J	S	O	I	G	A
G	N	I	T	H	C	A	Y	R	N	N	S
B	E	Z	B	A	L	L	T	S	G	I	T
O	G	N	I	C	N	E	F	E	A	M	I
X	R	A	C	K	E	T	T	C	D	M	C
I	W	E	I	G	H	T	S	I	F	I	S
N	O	L	H	T	A	I	R	T	C	W	R
G	J	U	D	O	T	E	N	N	I	S	P

ISBN: 9780170217095

6 How much will it cost to take the New Zealand team to London?

7 Research the importance of the venue, such as being either in northern/southern hemisphere or at high/low altitude, and how this affects participation.

8 Design a ticket for the London 2012 Games that could be used for either all sports or for a particular sport.

9 This diagram shows the three physical types and the typical body build of some Olympic athletes.

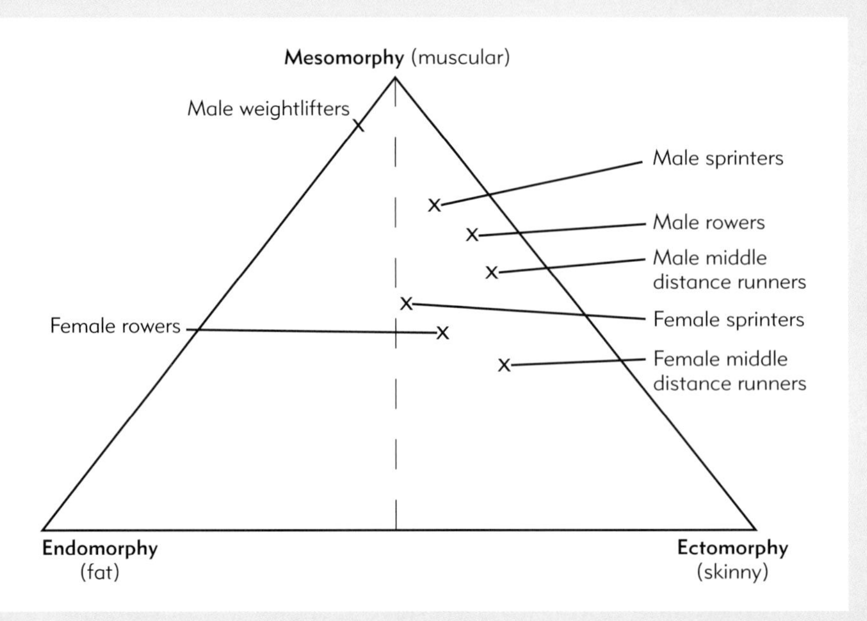

Research is the typical body build of other Olympic athletes. Where do you fit?

10 Hockey is played on a system of round robin pool play, with the winner of each pool playing the runner up of the other pool. The winner of these games play off for the final. If there are 12 men's teams (two pools) and ten women's teams (also two pools), design the tournament schedule for hockey for the Games. How many games of hockey would be played?

11 Research the relationship between speed, strength and endurance of various sports. Draw a triangle similar to that in question 9 above, and plot the sports you have researched onto it.

12 Make a collage of angles that athletes can make with their bodies.

13 Copy the grid of numbers below. Draw a continuous line connecting some of the numbers so that they add up to 1000, ending at 1000 in the bottom box.

Start ↓ (above the 58 column)

72	58	64	43	76
63	78	79	89	94
40	100	82	77	86
98	65	34	71	47
75	80	63	59	90
		1000		

ISBN: 9780170217095

14 London 2012

Make up at least ten problems that have an answer of 2012. You can use any combination of mathematical operations.

15 In the square puzzle below, each letter represents a value and the sum of the values for both rows and columns given. What value does each letter represent?

A	B	B	31
A	C	C	41
A	A	B	35
39	36	32	

16 One stand at the Olympic Stadium will hold 10 000 seats. If each number is made up of individual digit plates, how many number sevens are needed?

17 The athletic track at the Olympic Stadium has a circumference of 400 m and ten lanes. Construct a ten lane track, marking in the staggered starts for each event and the position of the 100 m line. Check to see that each athlete runs the correct distance for each event.

ISBN: 9780170217095

Answers

Exercise 2

1 The Olympic Games in this style was restarted in 1896, thus the reason for using the word modern, and the sports are summer sports.
2 Athens (Greece), Paris (France), St Louis (USA), London (England), Stockholm (Sweden), Antwerp (Belgium), Amsterdam (Netherlands), Los Angeles (USA), Berlin (Germany), Helsinki (Finland), Melbourne (Australia), Rome (Italy), Tokyo (Japan), Mexico City (Mexico), Munich (Germany), Montreal (Canada), Moscow (Soviet Union), Seoul (Korea), Barcelona (Spain), Atlanta (USA), Sydney (Australia), Beijing (People's Republic of China)
3 Two (Melbourne and Sydney)
4 Greater population, more established, closer to more countries.
5 17.4%
6 4 (18.2%)
8 First and Second World Wars.

Exercise 3

1 Games were founded in Olympia, Greece in 776 BC and we have continued to use the same symbols that were used at the time.
2 **a** VI **b** XX **c** XXXIII **d** LX **e** XCIX **f** LD
3 **a** 14 **b** 38 **c** 95 **d** 902 **e** 707 **f** 2000
4 **a** XXVI **b** XVI **c** IX **d** XXII **e** XVII **f** XXIX
5 **a** MCMXX **b** MCMXLVIII **c** MCMLXVIII **d** MCMXCII **e** MCMXII **f** MCMLXXXVIII
7 XXX
8 **a** X **b** XV **c** XLIII **d** LXI **e** CXV **f** CXXXIII

Exercise 4

2 **a** 18 363 km **b** 2504 km **c** 8153 km
d 11 hrs 35 min **e** 28 hrs 30 min **f** $588
g $4506 **h** $4332 **i** $4154
j **i** $22 754 **ii** $19 340.90 **iii** $2901.14
k **i** 16 c **ii** 20 c **iii** 86 c
l **i** 644 km/h **ii** 877 km/h **iii** 615 km/h
iv stopover

Exercise 5

1 **a** 12 hrs **b** 10 hrs **c** 1 hr **d** 8 hrs **e** 5 hrs **f** 3 hrs
2 **a** 9 p.m. **b** 7 p.m. **c** 10 a.m. **d** 5 p.m. **e** 4 a.m. **f** 12 p.m.
3 7.30 a.m., 28 July 2012.
4 **a** 8 p.m., 30 July **b** 7.50 p.m., 5 Aug
c 6.30 p.m., 5 Aug **d** 9 p.m., 4 Aug
e 9.30 p.m., 12 Aug **f** 5.30 p.m., 31 July
g 7 a.m., 3 Aug **h** 10.30 a.m., 4 Aug
i 8.00 a.m., 7 Aug **j** 8.30 p.m., 27 July
k 3.30 a.m., 13 Aug

Exercise 6

1 Water polo, basketball, hockey, athletics, swimming, handball, BMX, cycling, wheelchair tennis, diving, synchronised swimming, pentathlon, wheelchair rugby
2 **a** Football **b** Triathlon, marathon swimming
c Shooting **d** Beach volleyball
3 1200 m
4 800 m
5 2.5 km^2
6 **a** **i** ESE **ii** NW **iii** SW **iv** W **v** E
b Tourist sites
7 **a** £2660 **b** $5389.61
8 **a** 14–16 Craven Hill, Lancaster Gate, London
b $9860
c Any of the facilities listed
d Any of the points of interest listed
e London City

Exercise 8

1 2 3 4 5
6 7 8 9 10

12 **a** Five coloured rings intertwined symbolise the five continents.
c 1
d

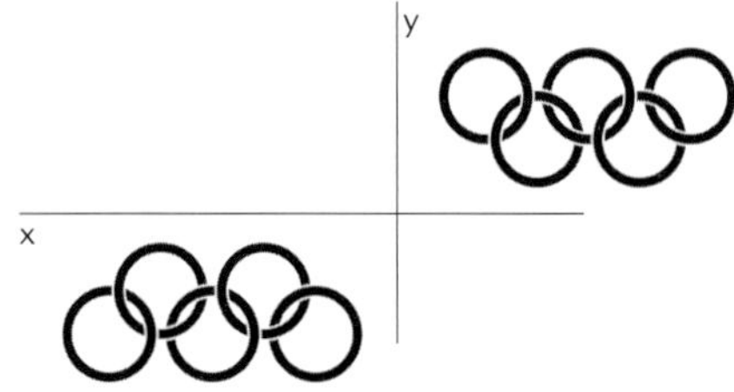

ISBN: 9780170217095

e i

Circles (C)	Intersection (I)
2	2
3	4
4	6
5	8
6	10
7	12
8	14
n	2n-2

ii Twice the number of circles subtract 2 is equal to the number of intersections (I = 2C-2).

f i 2
ii 2
iii 2
iv

Circles (C)	Least number of colours needed
1	1
3	2
5	2
7	2
9	2
n	2

Exercise 9

1 China
2 Australia, South Africa, Argentina, Colombia, Zimbabwe, Brazil, Peru, Venezuela
3 Any northern hemisphere country.
4

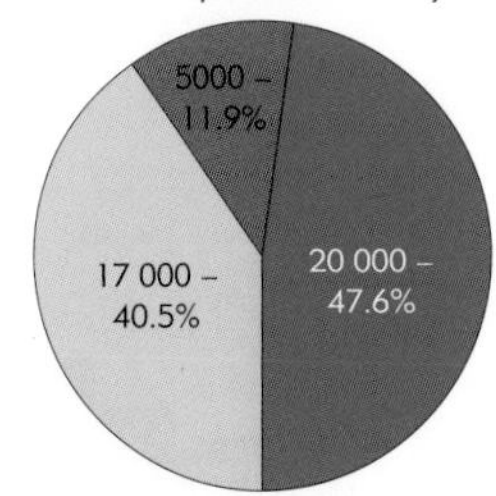

5 40.5%
6 **a** 6
b Union Jack on the flag
d Union Jack (Commonwealth) and Southern Cross (stars)
f

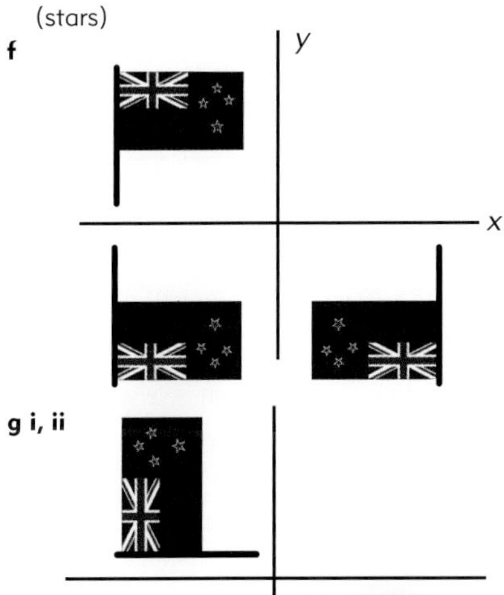

g i, ii

h i

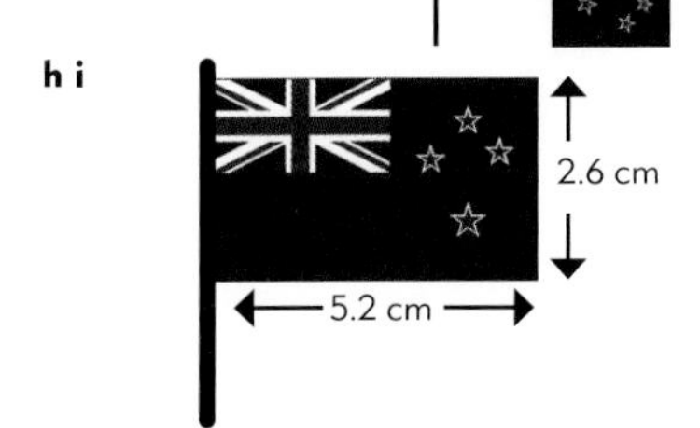

ii Scaled factor is half the original.

Exercise 10

1 Athletics
2 Gymnastics (trampoline)
3 5.3% (2 sports out of 38 – synchronized swimming and rhythmic gymnastics)
4 2/17
5 528
7 Mascots are created for children, connecting young people with sport. **Wenlock** is the name of a small town called Much Wenlock in Shropshire which hosted a precursor to the modern Olympic Games in the 19th century and Stoke **Mandeville** hospital in Buckinghamshire was the birthplace of the Paralympic Games.
8 **a** \$2325.58 **b** \$190 697 674 **c** 79%
d i 7140 m^2 **ii** 346 m
e i \$690 000 000 **ii** 6.9×10^8
9 **a** 9 000 000 **b** £1 350 000 000
10 1.13×10^7
11 **a** 23°C **b** 60 mm **c** 12-23°C
d i

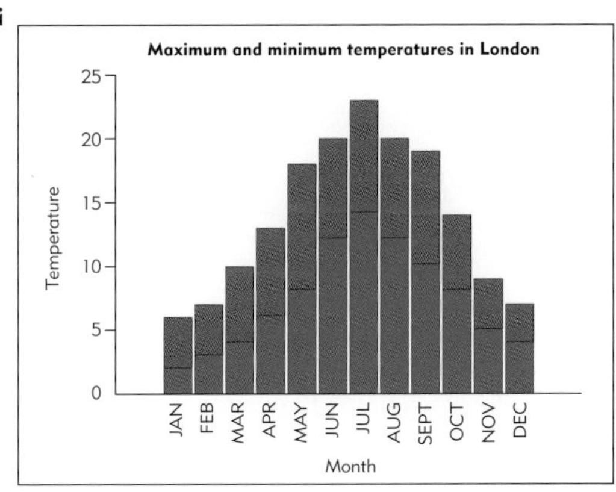

ii Maximum range 9°, minimum range 3°. Difference between summer and winter about 10°C
12 **a** 21/20 000 **b** 32/2 000
13 **a** Gymnastics and volleyball **b** Rowing
c 10 **d** Triathlon **e** 11 August

Exercise 11

1 9
2 0.9 km
3 1.6 km
4 **a** 9.7 sec
b i 100/9.7 = 10.3 m/s **ii** 10.3 x 3.6 = 37.1 km/hr
5 200 m by 0.0004 sec/m
6 **a** 3¾ **b** 12.5
7 357 cm
8 1.01 sec
9 0.2 sec
10 0.29 sec
11 9.28 sec
12 1.05 sec
13 2 hours, 33 minutes and 59 seconds
14 1.2 sec
15 **a** 55.1 sec **b** 20.2%
16 4 hr 33 min 1 sec
17 **a** 23 cm **b** 16 cm/91 cm = 17.6%
18

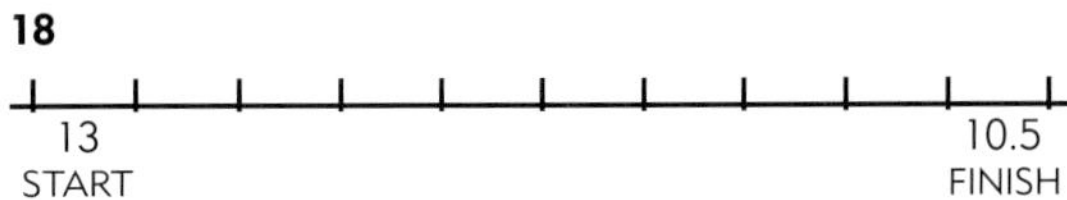

Exercise 12

1 86%
2 91 cm
3 5.7 cm
4 **a** 205, 206, 201, 205, 202, 203, 202, 197, 193, 192, 159 cm

ISBN: 9780170217095

b

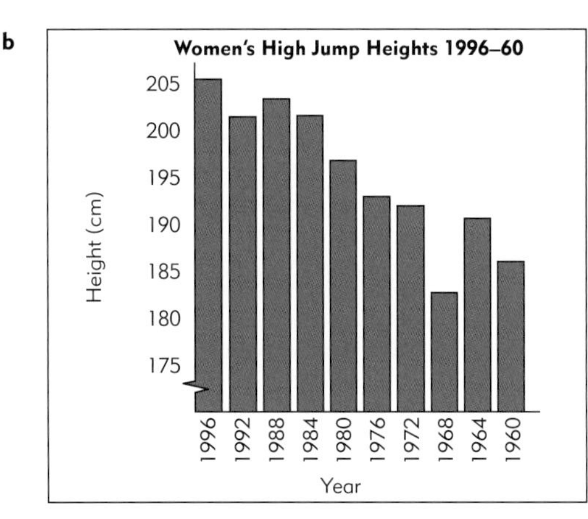

c Over ten years the height has increased 20 cm (average of 2 cm/year). There was a gradual increase except for 1968. The biggest increase was between 1968 and 1972.
d 29%

5 **a** No throw
b 1-20-96, 2-20-48, 3-20-09, 4-21-72, 5-21-96, 6-21-99
c 6 (1st), 5 (2nd), 4 (3rd), I (4th), 2 (5th), 3 (6th)

6 **a**

W WW
0 10 20 30 40 50 60 70 80 90 100
M M M M

b Men generally throw further than women, except in the discus.

7 **a** 1968 **b** 8.42 m **c** 0.70 cm/year
8 **a** 1996 **b** **i** 2.1 m **ii** 16.5 m

Exercise 13

1 **a** 20 min 12 sec **b** 42/2.1167 = 19.8 km/hr
c **i** 147 min **ii** 3 min 30 sec
2 **a** 3 min 5 sec **b** 2 hrs, 12 m and 11 s
3 **a** 210 min
b **i** 39 min 17 sec **ii** 78 min 34 sec
iii 117 min 51 sec
c 189 min **d** 1 hr 18 min 59 sec
e The Mall **f** Nelson's Column
g 3
h 42/1.609 = 26 miles 182 yards
i 16 min 42 sec

Exercise 14

1 **a** 8791, 8551
b 100 m, javelin, discus, long jump, shot put, 110 hurdles
c pole vault
3 900
4 **a** 827.5 **b** 516
5 **a** 6733, 6619
b 100 m hurdles, 200 m, 800 m, high jump
c javelin, shot put, long jump
6 **a** 7291/7 = 1041.6 points
b Natalie: 79.7 points per event, Hyleas: 96 points per event

Exercise 15

1 **a** 4 **b** 30
2 **a** 16 **b** 0.8 km
3 0.15 sec
4 **a** freestyle **b** 58.96 seconds
c breaststroke **d** freestyle
e 1.88 seconds **f** butterfly
5 **a** 4 min 3.84 sec **b** 8
c 30.48 sec
6 **a** 53.44
b Faster than 2000, 2004 time but 0.32 slower than 2008.
7 4.67 sec

Exercise 16

1 **a** 60 **b** 72.74
c 84.44 **d** 76.68
2

Comp.	Score	Placing
1	66.46	8
2	79.2	5
3	79.00	6
4	79.63	4
5	83.33	2
6	89.09	1
7	80.39	3
8	73.35	7

Exercise 17

1 Scale drawing length 25 cm, height 6 cm, width 0.5 cm, 6.2cm off the floor
2 **a** 9.575 **b** 8.92
3 1st – 9, 2nd – 5, 3rd equal – 4 and 10, 5th equal = 2 and 8
4 **a** China **b** 0.13 **c** Romania **d** 9.30
5 **a** Competitor 1: 6.5, 6.8, 6, 6, 7.7, Competitor 2: 7, 7.8, 7.5, 6, 6.9
b 33, 35.2
c Competitor 2

Exercise 18

1 The 82 kg lifter, because he was lighter in body weight
2 **a** 1, less faults **b** 2
c 1: 407.5 kg, 2: 410 kg, 3: 405 kg **d** 2, 1, 3 **e** 590 kg
3 **a**

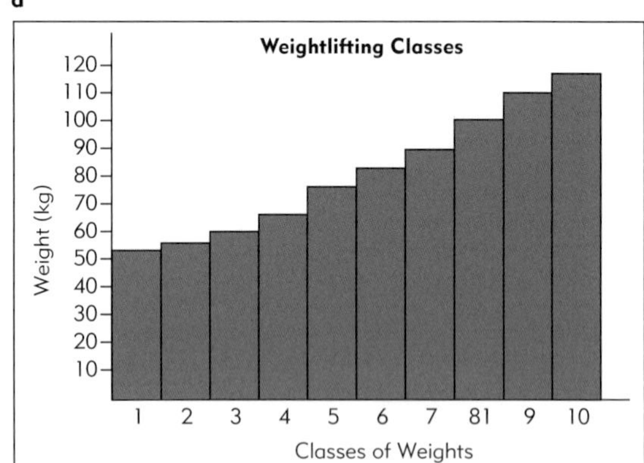

b 56 kg, 105 kg, >105 kg
c 22 lb
d 73:14
e 52 kg (Flyweight), 56 kg (Bantamweight), 60 kg (Featherweight), 67.5 kg (Lightweight), 75 kg (Middleweight), 82.5 kg (Light heavyweight), 90 kg (Middle heavyweight), 100 kg (First heavyweight), 110 kg (Second heavyweight), >110 kg (Super heavyweight)
4 Stones, logs
5 **a** 50 kg x 1, 20 kg x 1, 5 kg x 1, 1 kg x 2, 0.5 kg x 1 on each end + 20 kg barbell = $175 kg
b 50 kg x 1, 20 kg x 2, 5 kg x 1 on each end
c 50 kg x 1, 20 kg x 1, 10 kg x 1, 1 kg x 4, 0.5 kg x 1 on each end
d 20 kg x 1, 10 kg x 1, 5 kg x 1, 1 kg x 3, 0.5 kg x 1 on each end
6

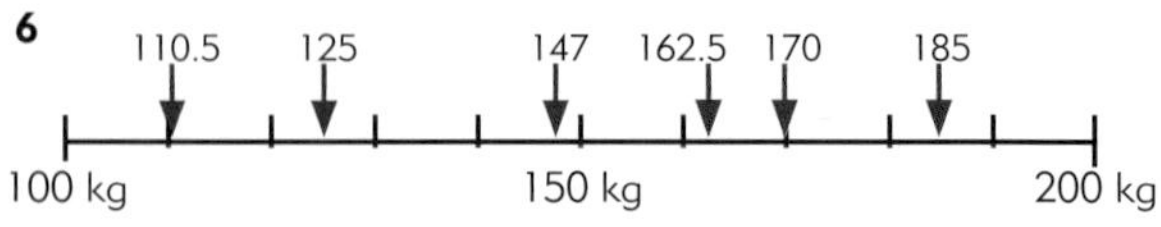

Exercise 19

1 **a** 3 **b** 17.6%
c Individual, more gold
d 1996: 1 gold, 1 silver, 4 bronze
e **i** Blyth Tait **ii** 5 **iii** 71.4%
f Mark Todd

ISBN: 9780170217095

2 a 8 b 7 c 0 d disqualification
e 8 f 18 g Competitor 3
h fastest time
3 a 42.9, 39.4, 45.8, 42.6, 39.2, 46.1, 42.6, 39.1, 46.3
b 128.1, 117.7, 138.2
c Competitor 3, 10.1 d 11.8

Exercise 20

1 a 1968–88 b Coxless fours
c 1972: 1 gold, 1 silver
d i 5 ii 2 rowers iii 9
e 5
f i 30 ii 22 iii 53
2 59.1 sec
3 a

	Men	Women
Single sculls	7.00	7.22
Double sculls	6.28	7.07
Coxless pairs	6.37	7.21
Coxed pairs	5.48	-
Coxless eights	6.07	-
Coxless fours	5.24	6.05
Coxed fours	5.41	6.16
Coxed eights	6.11	6.55

b

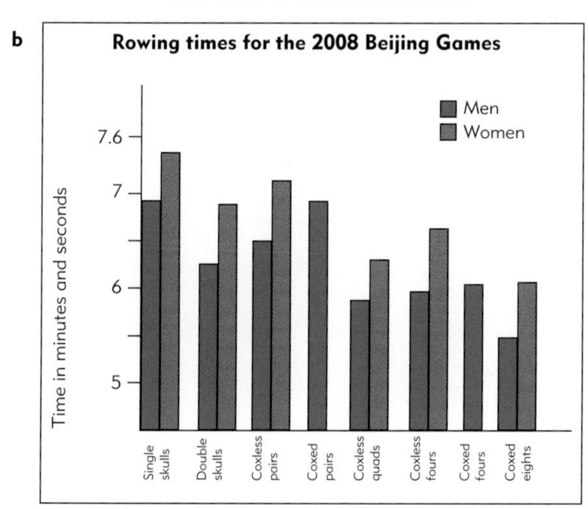

c Coxless pairs
d On average 33 seconds between men and women over all events. Eights is fastest for men and women. Women do not participate in coxed pairs or coxed fours.
4 a 2 b 4.8 m/s c i 87 sec ii 174 sec
5 a single: 6 m, doubles: 10.4 m, fours: 13.4 m, eights: 19.9 m
b 8 m c 40.2 %
6 a 378 kg b 794 kg
c 0.794 t d 48 kg
e 86.5 kg f 55 kg

Exercise 21

1 a i Kayak, one person, over 500 m
ii Kayak, two person, over 1000 m
iii Kayak, four person, over 1000 m
b 5
c Ian Ferguson
d 5
2 a 8.516 sec
b i 2:57.472 ii Yes
iii 14.337 sec
3 a 4 b 1992: 1 gold, 2 silver, 1 bronze
c 14
4 a 1 cm = 500 m b 22.75 km
c NE, ENE, SE, WSW, SE, WNW

d E from D
e From A to B
f D and E

Exercise 23

1 2008 Beijing
2 1414
3 89
4 6.3%
5 a 1984 b 1948, 1980
6 a 1912 b 66.67%
c 1908, 1912, 1920, 1924, 1952, 1936
7 a

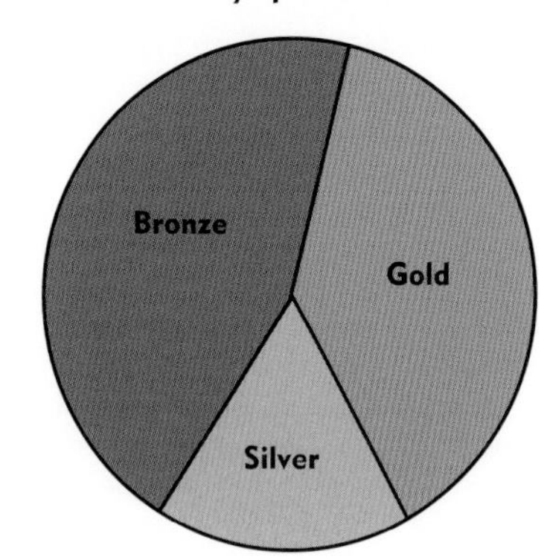

b Most medals gold. More than ⅓ medals are gold.
8 a 23 b 61 c 3.86 = 4
9 a Athletics
b i Athletics, yachting, rowing, equestrian, canoeing/swimming
ii Athletics, Rowing/Yachting, Equestrian, Canoeing, Swimming

c i

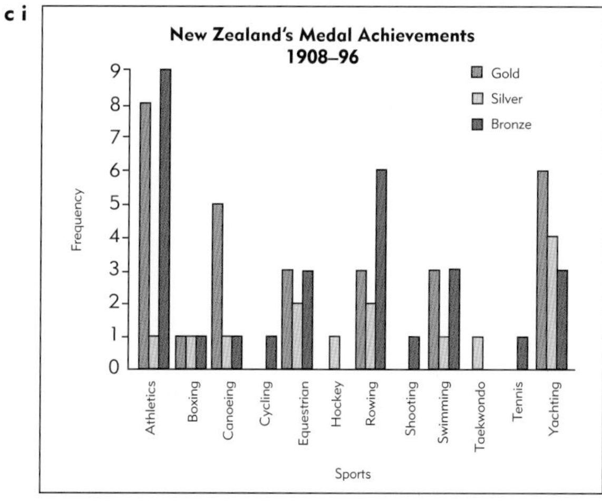

c ii Best sports are athletics, rowing, yachting and equestrian

Exercise 24

1 First and Second World Wars.
2 a 0%
b 42.4%. Participation by female athletes is increasing each year.
3 89.5% and 57.6%
4 7208
5 a 1469.2% b 3418.3%
6 a 54 b i 31 ii 23
7 a 1:175 000, 39:3 400 000, 3:1 375 000, 539:240 000 000, 17:81 250 000
b Australia, New Zealand, USA, Great Britain, People's Republic of China
c Australia d 97

ISBN: 9780170217095

e **i** 5:12, 56:195, 1:3, 152:539, 127:272
ii China

8 **a**

Games Funding 2012 London

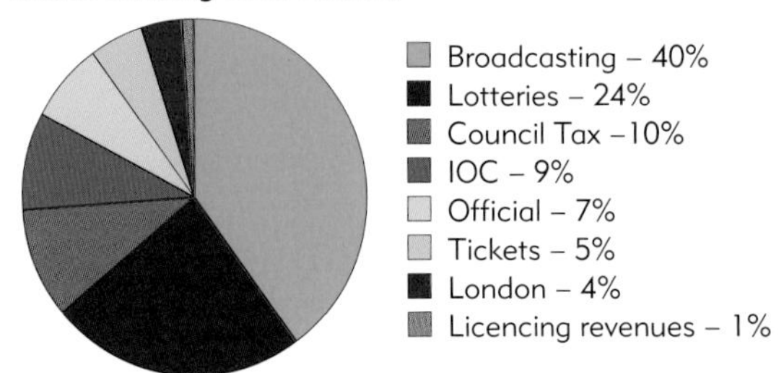

b Private donations
c **i** $1340 **ii** $440 **iii** $220
d stamps, coins, commemorative medals

9 **a** **i** 280% **ii** 843.8% **iii** 1476.9%
iv 14 900%
b Tickets available, venues, longer events and Games, easier travelling for spectators, more affluence.
c Athens 1896 = 62 athletes/day, London 2010 = 895 athletes/day – London will have greater participation.

10 **a** New Zealand is one of the first countries in the world to develop different-shaped stamps (squares, rectangles, circles, triangles). There is no maximum or minimum size except it must be suitable for an envelope. New Zealand's standard size is 25 mm x 30 mm.
c 40c, 80c, $1, $1.20, $1.50, $2
d

Stamps required	
$1.20	40c, 80c
$1.90	40c , $1.50
$4.50	3 x $1.50
$7.20	4 x $1.80
$10.00	$1.00, 5 x $1.80

11 **a** $12.50
b $5.10. $4.90 change.
c 69 550mm^2 $3.20

12 **a** £6.08
b **i** £5.92
ii $12.45
c London to NZ $1.59. NZ to London $1.90
d UK postage is cheaper.

Exercise 25

1 Mobility disabilities, amputees, paraplegics, limb deformities, cerebral palsy, visually impaired, dwarfism, multiple sclerosis, congential deformities etc.
2 ALA (Amputee and Les Autres), CP (Cerebral Palsy), ID (Intellectual Disability), VI (Visually Impaired), WC (Wheelchair)
4 After the Olympics every four years.
5 Yes
6 1960 at Rome Olympics, 400 athletes from 23 countries.
7 Neroli Fairhall in archery at the 1984 Los Angeles Games.
8 Trisha Zorn won 55 medals – 41 of which were gold – between 1980–2004, in swimming.

Exercise 26

1 'One World, One Dream'
2 Started in Tian Tan Gong Yuan (Temple of Heaven) in the south of the city, through the city to the National Stadium in the north of the city for the opening ceremony on 6 September 2008.

3

4 Archery, athletics, boccia, road cycling, track cycling, equestrian, football (five-a-side and seven-a-side), goalball, judo, powerlifting, rowing, shooting, swimming, table tennis, volleyball, wheelchair basketball, wheelchair fencing, wheelchair rugby, wheelchair tennis.
5 Athletics, boccia, powerlifting, shooting, swimming, wheelchair rugby, cycling.
6 Oscar Pistorius is a double amputee, runs on blades and competes in 100 m, 200 m and 400 m athletics.
7 Natalie du Toit in swimming.
8 13 year old Eleanor Simmonds of Great Britain, in the 100 m S6 freestyle event.
9 **b** Brazil, South Korea, Portugal, Great Britain and Hong Kong.

10 **a** **i** 104 **ii** 62 **iii** 32 **iv** 26
b **i** 11:70 **ii** 3:56 **iii** 9:136 **iv** 11:47
c cycling
d **i** 54.5% **ii** 90.9% **iii** 45.5% **iv** 27.3%
e **i** 69/280 **ii** 2/15 **iii** 11/175 **iv** 4/175

11 **a** **i** 211:332 **ii** 79:170 **iii** 10:21 **iv** 2:5
b **i** 21:106 **ii** 19:143 **iii** 1:6 **iv** 8:81
c South Africa
d **i** 18.8% **ii** 8.9%

Exercise 27

1 29 August to 9 September 2012.
2 Archery, athletics, boccia, road cycling, track cycling, equestrian, football (five-a-side and seven-a-side), goalball, judo, powerlifting, rowing, shooting, swimming, table tennis, volleyball, wheelchair basketball, wheelchair fencing, wheelchair rugby, wheelchair tennis (20 sports)
3 Two million
4 140
7 Intellectual disabilities

Putting it into practice

AS 91030

1	**a** 1000	**b** 0.05695 km	**c** 650,000 cm
2	**a** 5000	**b** 1.234 kg	**c** 0.453g
3	**a** 225 mins	**b** 150 sec	**c** 2 hr 47 min
4	**a** 10 m/secs	**b** 2 hr 27min	**c** male by 1:33:8 sec
5	**a** 3260 g	**b** 4.4 lb	**c** 182.1 mm
6	**a** 15 cm/5.9 in	**b** 0.7874 in	**c** 2.39 m/s, 2.15 m/s
7	**a** **i** 12 m	**ii** 9.5 m	**iii** 8.5 m
	b **i** 62 kg	**ii** 57 kg	**c** 46 kg
8	**a** 750 m^3	**b** 750 000L	**c** 687.5 m^3
9	**a** 15.97 lb	**b** 3.58 m^2	**c** 9.7%
10	**a** 9.14 m	**b** 8.5 m	**c** 30.79 km/h
11	**a** 4.57 m	**b** 17.6 m	**c** 5.28 m
12	**a** 10%	**b** -4.1%	**c** $14 800 000

13 Discuss with your teacher.

ISBN: 9780170217095

AS 91034

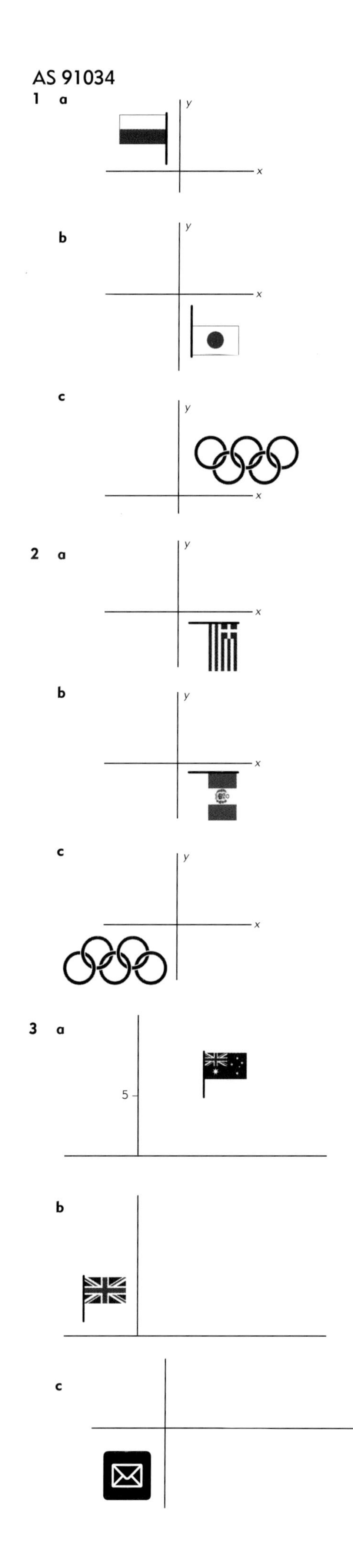

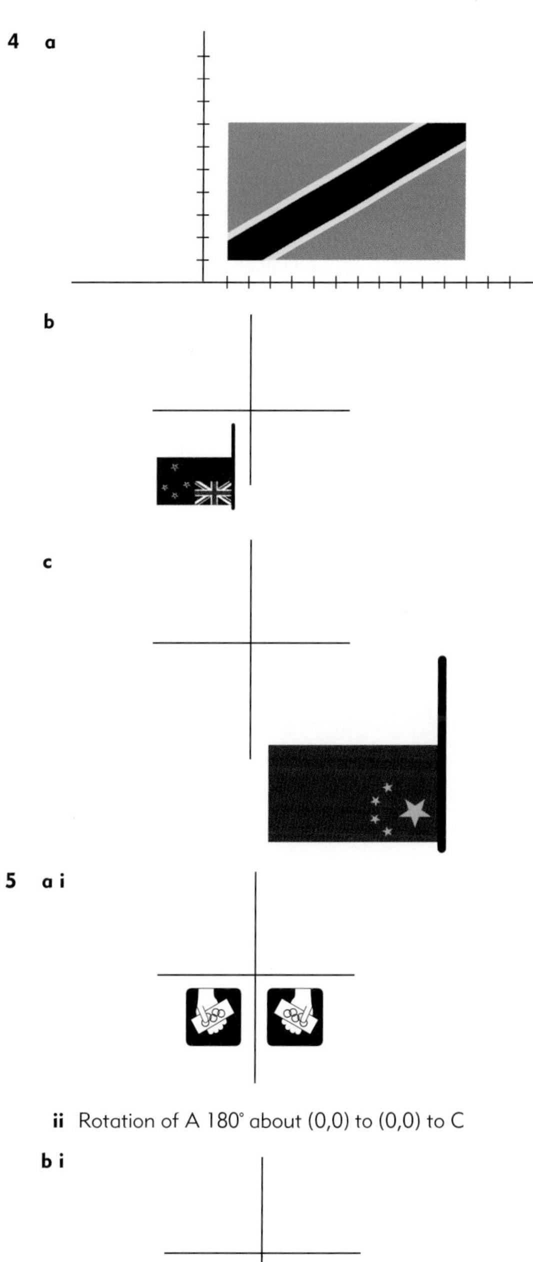

5 a i

ii Rotation of A 180° about (0,0) to (0,0) to C

b i

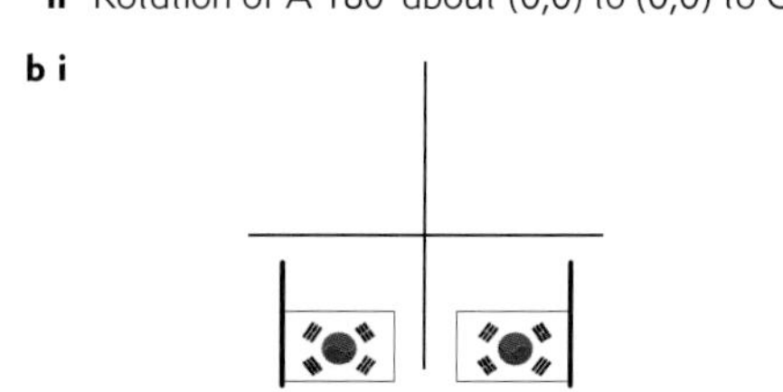

ii Reflect in x axis.

c i

ii Rotate about (0,0) and enlarge by a scale factor of 2 or, reflex in the line y=x and enlarge by a scale factor of 2.

ISBN: 9780170217095

6 **a i**

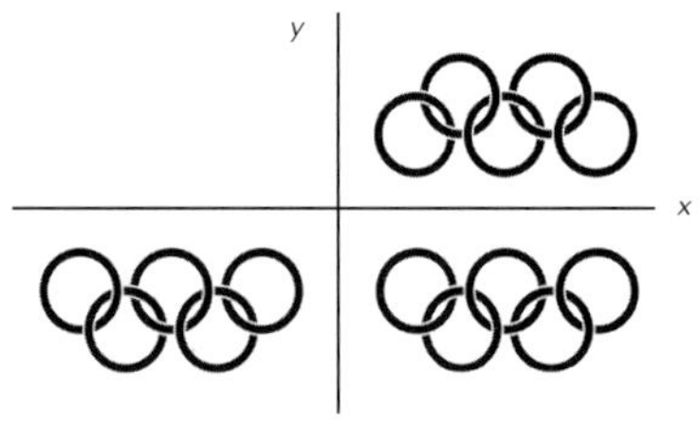

ii Reflection in x-axis and reflection in the y-axis.

7 **a** Reflection in y-axis.
b Rotation 180° about (0,0)
c A to B – reflection in x axis
B to C – reflection in y axis

8 **a** Rotation about (0,0) 180°
b Enlarge by scale factor of 2 using (0,0) then rotate 180°.
c Rotate 180° about (0,0) to B then reflect B to C in x axis.

Enrichment

Word search

5

C	O	O	C	Y	C	L	I	N	G	R	G
A	R	C	H	E	R	Y	O	T	R	A	Y
N	T	A	E	K	W	O	N	D	O	O	M
O	U	A	V	C	Q	Z	U	H	W	P	N
E	P	S	T	O	R	J	S	O	I	G	A
G	N	I	T	H	C	A	Y	R	N	N	S
B	E	Z	B	A	L	L	T	S	G	I	T
O	G	N	I	C	N	E	F	E	A	M	I
X	R	A	C	K	E	T	T	C	D	M	C
I	W	E	I	G	H	T	S	I	F	I	S
N	O	L	H	T	A	I	R	T	C	W	R
G	J	U	D	O	T	E	N	N	I	S	P

10 Men: 31 games. Women: 23 games

Men

Pool A

A v B A v D A v F A v C A v E
C vD B v E B v D B v F B v C
E v F C v F C v E C v D F v D

Winner Pool A v
Winner Pool B

Pool B

G v H G v J G v L G v I G v K
I v J H v K H v J H v L H v I
K v L I v L I v K I v J L v J

Winner Pool B v
Winner Pool A v

Winner of Pool A
Play-off
Winner of Pool B
Play-off
Winner

Women

Pool A

A v B A v C A v D A v E B v D
C v D D v E B v E B v C C v E
E B C D A

Pool B

F v G F v H F v I F v J G v I
H v I I v J G v J G v H H v J
J G H I F

Winner of Pool A V Runner-up of Pool B
Winner of Pool B v Runner-up of Pool A

Winner of Pool A
v
Winner of Pool B

Winner

13

	58		43	
		79		94
	100	82	77	86
98	65			
75	80	63		
		1000		

15 A = 13, B = 9, C = 14

16 4000

ISBN: 9780170217095